COSMIC PLASMA

ASTROPHYSICS AND SPACE SCIENCE LIBRARY

A SERIES OF BOOKS ON THE RECENT DEVELOPMENTS
OF SPACE SCIENCE AND OF GENERAL GEOPHYSICS AND ASTROPHYSICS
PUBLISHED IN CONNECTION WITH THE JOURNAL
SPACE SCIENCE REVIEWS

VOLUME 82

COSMIC PLASMA

by

HANNES ALFVÉN

University of California, San Diego, La Jolla, California

and

Royal Institute of Technology, Department of Plasma Physics,
Stockholm, Sweden

D. REIDEL PUBLISHING COMPANY

DORDRECHT : HOLLAND / BOSTON : U.S.A.

LONDON : ENGLAND

Library of Congress Cataloging in Publication Data

CIP

Alfvén, Hannes, 1908–
Cosmic plasma.

(Astrophysics and space science library ; v. 82)
Bibliography: p.
Includes index.
1. Plasma (Ionized gases) 2. Outer space.
3. Cosmology. I. Title. II. Series.
QB500.A39 523.01 80-27868
ISBN 90-277-1151-8

Published by D. Reidel Publishing Company,
P.O. Box 17, 3300 AA Dordrecht, Holland.

Sold and distributed in the U.S.A. and Canada
by Kluwer Boston Inc.,
190 Old Derby Street, Hingham, MA 02043, U.S.A.

In all other countries, sold and distributed
by Kluwer Academic Publishers Group,
P.O. Box 322, 3300 AH Dordrecht, Holland.

D. Reidel Publishing Company is a member of the Kluwer Group.

Printed in The Netherlands

TABLE OF CONTENTS

PREFACE

The general background of this monograph and the aim of it is described in detail in Chapter I. As stated in I.7 it is written according to the principle that "when rigour appears to conflict with simplicity, simplicity is given preference", which means that it is intended for a rather broad public. Not only graduate students but also advanced undergraduates should be able to understand at least most of it.

This monograph is the result of many years of inspiring discussions with a number of colleagues, for which I want to thank them very much. Especially I should mention the groups in Stockholm and La Jolla: in Stockholm, Dr Carl-Gunne Fälthammar and many of his collaborators, including Drs Lars Block, Per Carlqvist, Lennart Lindberg, Michael Raadu, Staffan Torvén, Miroslav Babić, and Ingvar Axnäs, and further, Drs Bo Lehnert and Björn Bonnevier, all at the Royal Institute of Technology. Of other colleagues in Sweden, I should mention Dr Bertel Laurent, Stockholm University, Dr Aina Elvius, The Stockholm Observatory, and Dr Bengt Hultqvist, Kiruna.

In La Jolla my thanks go first of all to Dr Gustaf Arrhenius, who once invited me to La Jolla, which was the start of a most interesting collaboration; further, to Dr W. B. Thompson, who has offered a challenging and inspiring criticism, and to Dr Asoka Mendis, with whom I have had almost daily discussions for a number of years. I have had a rewarding collaboration with members of Dr Carl McIlwain's group, especially with Dr Elden Whipple, and Dr Walker Fillius; and further, with Drs Margaret Burbidge, H. E. Smith, and A. Marsher.

Of colleagues outside La Jolla, I am indebted to the space research group at Rice University under Alex Dessler, the group directed by Dr T. Krimigis at Johns Hopkins University, Dr Syun-Ichi Akasofu at the University of Alaska, Dr G. de Vaucouleurs and Dr Walter Heikkila at the University of Texas, and Dr H. Arp at the California Institute of Technology.

I am grateful to Mrs Jane Chamberlin, and Messrs Max Marconi and Dave Cordova for helping me with the manuscript and to Mr Johan Bonde and Mrs Margareta Malmort for the final editing. Mrs Eva Florman, Kaj Forsberg, Carine Hamrin, Siri Lilliesköld and Kerstin Vikbladh have given me much appreciated assistance.

The work has been supported in the United States by a number of grants from NASA and NSF and in Sweden by Naturvetenskapliga Forskningsrådet. Professor Carl-Gunne Fälthammar has kindly placed the resources of his institute at my disposal.

CHAPTER I

SURVEY

I.1. Experimental and Theoretical Approach to Plasma Physics

Plasma physics started along two parallel lines. One of them was the hundred-year-old investigations into what was called 'electrical discharges in gases'. To a high degree, this approach was experimental and phenomenological, and only very slowly did it reach some degree of theoretical sophistication. Most theoretical physicists looked down on this field which was complicated and awkward. The plasma exhibited striations, double layers, and an assortment of oscillations and instabilities. The electron temperature was often found to be one or two orders of magnitude larger than the gas temperature, with the ion temperature intermediate. In short, it was a field which was not well suited for mathematically elegant theories.

The other approach came from the highly developed kinetic theory of ordinary gases. It was thought that, with a limited amount of work, this field could be extended to include ionized gases. The theories were mathematically elegant and claimed to derive all of the properties of a plasma from first principles. In reality, this was not true. Because of the complexity of the problem, a number of approximations were necessary which were not always appropriate. The theories had very little contact with experimental plasma physics; all awkward and complicated phenomena which had been observed in the study of discharges in gases were simply neglected.

In cosmic plasma physics, the experimental approach was initiated by Birkeland, who was the first one to try to connect laboratory plasma physics and cosmic plasma physics. (Neither term was used at that time!) Birkeland observed aurorae and magnetic storms in nature and tried to understand them through his famous terrella experiments (Birkeland, 1908). He immersed his terrella in plasma and found that under certain conditions luminous rings were produced around the poles (see Figure I.1). Birkeland identified these rings with the auroral zones. As we know today, this is essentially correct. Further, he constructed a model of the polar magnetic storms, assuming that the auroral electrojet was closed through vertical currents (along the magnetic field lines). This was a natural assumption because his experiment was a study of electric current in 'vacuum'. This idea was also correct as a first approximation (the current system is more complicated than he thought). Hence, although Birkeland could not have known very much about the detailed structure of the magnetosphere, research today essentially follows Birkeland's lines, especially in the respect that the contact between laboratory experiments and astrophysics is important (Dessler, 1979; Potemra, 1978).

Unfortunately, the progress along these lines was disrupted. Theories about plasmas, at that time called ionized gases, were developed without any contact with laboratory plasma work. In spite of this – or perhaps because of this – belief in the theories was so strong that they were applied directly to space. One of the results was the Chapman – Ferraro theory (for a review, see Akasofu and Chapman, 1972) which became accepted to

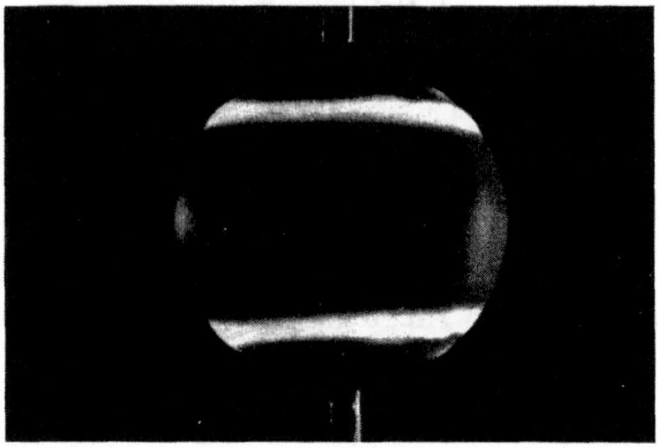

Fig. I.1. Birkeland's terrella experiment (a modern version) demonstrating what happens with a magnetized body when immersed in a plasma (under rather general conditions). The plasma penetrates to the auroral zones. (*L. Block*)

such an extent that Birkeland's approach was almost completely forgotten. For thirty or forty years, Birkeland's results were often ignored in textbooks and surveys, and all attempts to revive and develop them were neglected.

The crushing victory of the theoretical approach over the experimental approach lasted only until the theory was used to make experimentally verifiable predictions. From the theory, it was concluded that in the laboratory, plasmas could easily be confined in magnetic fields and heated to such temperatures as to make thermonuclear release of energy possible. When attempts were made to construct thermonuclear reactors, a confrontation between the theories and reality was unavoidable. The result was catastrophic. Although the theories were generally accepted, the plasma itself refused to believe in them. Instead, the plasma showed a large number of important effects which were not included in the theory. It was slowly realized that one had to develop new theories, but this time in close contact with experiments.

The 'thermonuclear crisis' did not affect cosmic plasma physics very much. The development of the theories continued because they largely dealt with phenomena in regions of space where no real check was possible. The fact that the basis of several of the theories had been proved to be false in the laboratory had very little effect. One said that this did not necessarily prove that they must also be false in the cosmos! Much work was done in developing these theories, leading to a gigantic structure of speculative theories which had no empirical support.

Another confrontation occurred when space missions made the magnetosphere and interplanetary space accessible to physical instruments. The first results were interpreted in terms of the generally accepted theories or new theories built on the same basis. However, as observational techniques advanced, it became obvious that several of these theories were not applicable. The plasma in space turned out to be just as complicated as laboratory plasmas and to follow the same basic laws. Those theories which had little contact with experimental results were now finally disproved by observational facts. Our picture of space around the Earth has changed radically in the past decades, as illustrated in Figure I.2.

Presently, our knowledge of laboratory plasmas is based on very sophisticated

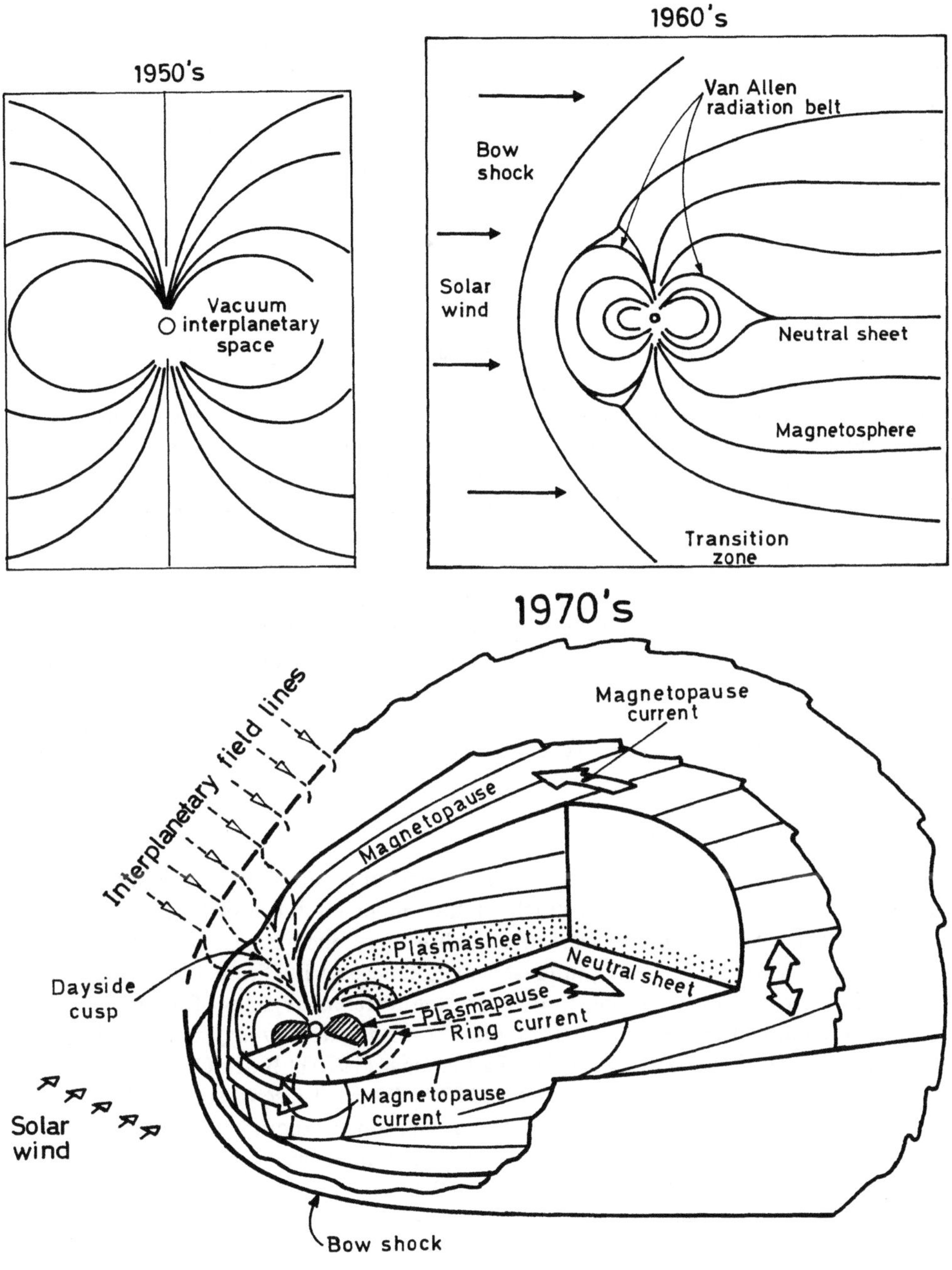

Fig. I.2. (a) Up to the beginning of the space age it was generally assumed that the Earth was surrounded by vacuum and its dipole field was unperturbed (except at magnetic storms). (b) The first space measurements showed the existence of the Van Allen belts, the magnetopause, the neutral sheet in the tail and the bow shock. (c) New measurements have made the magnetic field description increasingly complicated. The electric field description leads to a new picture, illustrated in Figure III.14.

'diagnostics' requiring magnetic and electric probes, advanced laser probes, etc. Our knowledge of magnetospheric plasmas is based on comparably sophisticated diagnostics by *in situ* measurements in space. However, these measurements are necessarily confined to those regions of space which are accessible to spacecrafts. As the hope of making *in situ* measurements in more distant regions is bleak, we may ask to what extent it is possible to understand these distant plasmas. Much of this volume will be devoted to this question.

I.2. Plasma Phenomena in Laboratory and Space

There are good reasons to believe that in several important respects, the basic properties of laboratory plasmas and space plasmas are the same. But as the linear scale of laboratory experiments is say 0.1 m, a scaling by a factor of 10^9 to 10^{12} is necessary in order to apply the laboratory results to planetary magnetospheres or to the solar magnetosphere (i.e., the heliosphere or interplanetary space). Furthermore, to apply the results to galactic, including cosmological problems, we have to scale by another factor of 10^9 to 10^{12} or more. As different plasma parameters obey different scaling laws, translation from one region to another is often difficult.

Figure I.3 gives a survey of some important plasma phenomena. Several of them are probably related, but a scaling by a very large factor is essential in order to check this. Hence, important progress can be expected if we can translate knowledge gained from measurements of laboratory and magnetospheric plasmas to more distant regions. However, this can be done only if we solve the very difficult problem of scaling plasma phenomena. We shall now give a survey of the scaling problems.

I.2.1. SCALING PROCEDURES

The advance of thermonuclear research has resulted in a better understanding of plasma phenomena in the laboratory and much of this understanding can be applied to space problems. However, thermonuclear investigations are increasingly focused on those problems which are of technical interest and not necessarily those which are of most interest in cosmic physics. Hence, to compare the different regimes of plasma physics, it is essential to design *laboratory experiments which simulate cosmic problems*. Surveys of such experiments have been given by Fälthammar (1974), Block (1976), and in the Nagoya University Research reports (1977, 1979). We shall discuss some of these experiments in Chapter II. Further, it is now possible to perform experiments in space.

Basic laws for scaling have been known for some time (summarized, e.g., in Alfvén and Fälthammar, 1963 (henceforth referred to as CE); recent progress is surveyed by Siscoe, 1979). However, it should be remembered that scaling laws depend on a theoretical formalism which is probably only applicable to passive plasma regions (II.9.1), whereas it has not yet been clarified how scaling between active plasma regions should be done.

Predictions of what could be expected from the first space mission to a planet have so far not been very successful. The predictions are, of course, based on generally accepted scaling theories, and almost every new space mission since has demonstrated that our understanding of space plasma is still not very good.

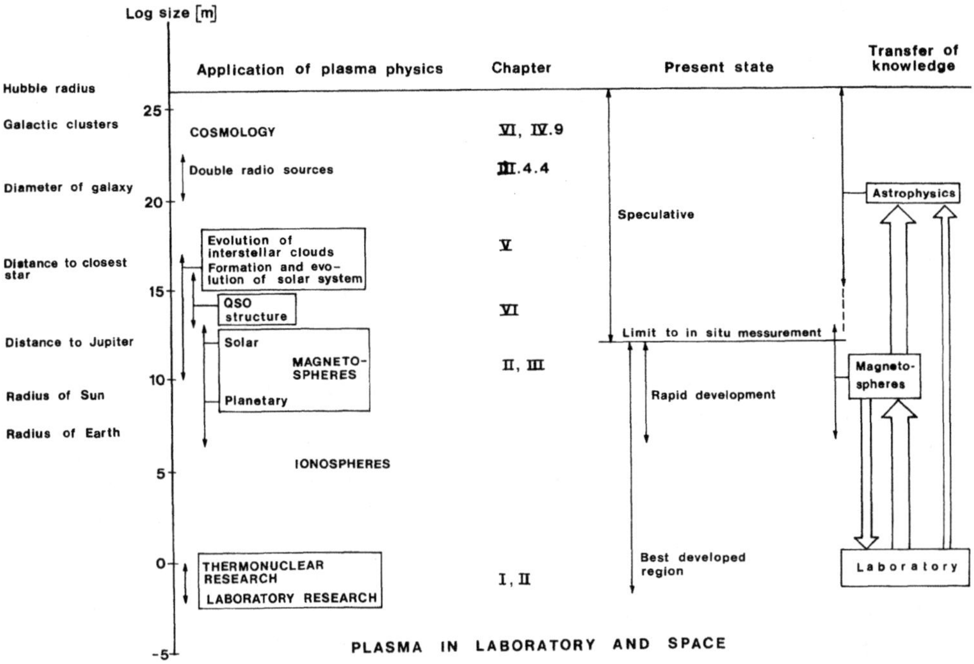

Fig. I.3. Plasmas in laboratory and in cosmos. The diagram gives a survey of the size of the regions which are of interest, and the chapters in which they are treated. It also indicates the limit between the regions which can be explored by high quality diagnostics and those which cannot. Hence, the latter must necessarily be regarded as speculative. The transfer of knowledge gained from laboratory experiments to magnetosphere research is now supplemented by a transfer in the opposite direction. Knowledge of more distant regions will also be obtained in this way.

I.2.2. SIMULATION EXPERIMENTS

It was first thought that one could design laboratory experiments which gave an adequate simulation of space conditions. This is possible only in rare cases because the plasma state depends on many parameters which often scale in different ways. To clarify cosmic plasma phenomena by means of laboratory experiments is very complicated. There are basically two types of simulation which have turned out to be fruitful.

One is *'pattern simulation'*. An experimental apparatus is constructed with a geometrical configuration which simulates space conditions. (The Birkeland experiment is an example of this.) However, it should be observed that the plasma parameters must be chosen according to certain criteria, so that the laboratory pattern really is relevant to the cosmic situation.

The second one is sometimes called *'process simulation'*. This essentially means that we explore the basic behavior of a plasma in the laboratory. Such experiments are important in building the theoretical foundation of plasma physics in general. They have shown that many of the conclusions which were drawn from classical plasma theory were wrong, and once again demonstrated that science is basically empirical. Theory is of value only when developed in close contact with reality.

Besides laboratory simulation, *computer simulation* is also widely used. This can be a very valuable substitute for experiments. However, it can never replace process simulation because the computer input must include the basic properties of a plasma which can be found only by experiments (in laboratory or in space). The computer is not a good physicist if its programmer is not.

I.2.3. COMPARISON BETWEEN LABORATORY AND SPACE INVESTIGATIONS

Laboratory experiments are, in several respects, easier to do than investigations by *in situ* measurements in space. Usually, the latter measurements can be made only in an orbit-governed space–time sequence which is often far from optimum. Even within these limits, the cost of space missions imposes further serious constraints. The advent of active experiments in space is an important step towards a better understanding of the real space plasma. However, the fruitfulness of this activity should be greatly enhanced if combined with suitably designed laboratory experiments.

The relevance of laboratory plasma experiments is not limited to simulation. In the past, essential concepts such as electric fields along magnetic field lines, electrostatic particle acceleration in double layers (II.6), current filamentation (II.4) and the general tendency of a plasma to be non-homogeneous were found in the laboratory without inspiration from space research. These ideas were later transferred to space problems. In the future, it can be expected that progress in space plasma physics can benefit from laboratory plasma experiments in a much broader sense than now.

I.3. Field and Particle Aspects of Plasmas

We have discussed the translation of results from one region to another. However, another translation which is equally important is the translation between a magnetic field description and a current description of plasma phenomena. Space measurements of magnetic fields are relatively easy, whereas direct measurements of electric currents are very difficult, in many cases impossible. Hence, it is natural to present the results of space exploration (from spacecrafts and from astrophysical observations) with pictures of the magnetic field configuration. Furthermore, in magnetohydrodynamic theories, it is convenient to eliminate the current ($\mathbf{i}$ = current density) and to represent electric currents by curl $\mathbf{B}$. This method is acceptable in the treatment of a number of phenomena (see Figure I.4).

However, there are also a number of phenomena which cannot be treated in this way, but which require an approach in which the *electric current is taken account of explicitly*. The translation between the magnetic field description and the electric current description is made with the help of Maxwell's first equation

$$\nabla \times \mathbf{B} = \mu_0 \left(\mathbf{i} + \frac{\partial \mathbf{D}}{\partial t} \right) \tag{1}$$

The displacement current can usually be neglected. However, it is often convenient to account for the kinetic energy of a magnetized plasma by introducing the permittivity $\epsilon = \epsilon_0 (1 + (c/V_A)^2)$, where c and V_A are the velocities of light and of hydromagnetic waves (CE 3.4.4). If this is done, the displacement current is often large.

As is shown in Figure I.4, some types of plasma phenomena should be described by

TRANSLATION FORMULA

$$\nabla \times \mathbf{B} = \mu_0 \mathbf{i}$$

MAGNETIC FIELD DESCRIPTION

Magnetic fields are:
Measured rather easily
Basic for plasma *anisotropy*
including high energy particle motion
Gives a good description of some
waves in plasmas

ELECTRIC CURRENT DESCRIPTION

Electric currents are:
Difficult to measure directly but
Essential for understanding:
Double layers
Transfer of energy from one region
to another
Current sheet discontinuities
Cellular structure of space
Magnetic substorms, solar flares

Applicable in

Passive plasma regions | Active plasma regions

The plasma dualism is somewhat analogous to the general particle-field dualism in physics.

The current description requires a new formalism with ELECTRIC CIRCUITS as an important ingredient.

Fig. I.4. Dualism in plasma physics.

the field picture and others by the current picture. Attempts to describe a certain group of phenomena with the wrong formalism often lead to erroneous results. This will be discussed in some detail in Chapter II.

Phenomena which cannot be understood without explicitly accounting for the current are:

(A) Energy transfer from one region to another (see II and III).
(B) Formation of double layers (see II.6).
(C) The occurrence of explosive events such as solar flares (III.8), magnetic substorms (III.5.1), possibly also 'internal ionization' phenomena in comets (Wurm, 1963; Mendis, 1978) and stellar flares.
(D) Double layer violation of the Ferraro corotation (III.3). Establishing 'partial corotation' is essential for the understanding of some features of the solar system (see ESS, 17, 18).
(E) Formation of filaments in the ionospheric aurora, the solar atmosphere, the nebulae and in the ionosphere of Venus (II.4 and II.9.3).
(F) Formation of current sheets which may give space a 'cellular structure' II.10 and VI).

Exploration of those plasma properties which can be described by the magnetic field concept have in general been successful. However, this is not the case for those phenomena which cannot be understood by this approach. The present monograph shall concentrate on the latter, and try to give *a survey of cosmic plasmas based on the particle (electric current) aspect.*

I.4. Present State of the Classical Theory

The classical theory, the foundation of which is due to Chapman and Cowling, has been successful in accounting for those phenomena which, according to Figure I.4, are related to the field description. The propagation of waves in plasmas belongs to this category as does the motion of high-energy charged particles and the drift of low-energy particles, including mirror effects. However, a real plasma has many properties which are not easy to describe by the classical formalism. Certainly, it is possible to produce externally heated plasmas (e.g., cesium-plasma) which are completely quiescent, and in these, the classical formalism seems to give a good description (for references see, for example, Motley, (1975).

However, as soon as an *electric current* is passed through a quiescent plasma, a number of complicated phenomena are produced which require an extensive development of classical theory, sometimes even a new approach (see IV).

The most important of these 'anomalous' properties are:

(a) Sometimes the plasma becomes more '*noisy*' than theoretically expected.

(b) The *energy distribution* becomes strongly *non-Maxwellian*; there is a considerable, sometimes extremely large, excess of high energy particles. The velocity distribution is very often highly anistropic.

(c) The *electron temperature* may be orders of magnitude larger than the *ion temperature*, which may be substantially higher than the *neutral gas temperature*, which again may differ from the temperature of the electrodes and the walls of the discharge tube or the temperature of the dust (in case the plasma is 'dusty', see IV.7). Certainly this can be accounted for by classical theory; but astrophysical plasmas are often treated as if the mentioned temperatures were necessarily equal.

Further, there are a very large number of phenomena which often are referred to as 'instabilities' (Krall and Trivelpiece, 1973). Examples of such phenomena are:

(d) At sufficiently *large current density*, the plasma may contract into *filaments* (II.4).

(e) In case the plasma consists of a *gas mixture*, the components often *separate* (IV.3).

(f) When the electron drift velocity exceeds the thermal velocity, *double layers* may be produced. Under certain conditions, these may explode and produce violent current surges which lead to the acceleration of plasma particles, sometimes as high as cosmic ray energies (II.6 and IV.10).

(g) It is also possible that at current densities above a certain limit, the resisitivity of the plasma increases by orders of magnitude (*anomalous resistivity*) but arguments for the occurrence of such a phenomena are still not convincing. (In several cases, the formation of double layers may have been interpreted as anomalous resisitivty.)

(h) When β (i.e., the ratio between gas pressure and magnetostatic pressure) approaches unity, the geometrical configuration of a plasma may be deformed. The long list of such instabilities includes the *kink instability* and the *sausage instability*.

I.5. Boundary Conditions. Circuit Dependence

In applying the classical theory and its modern development, the importance of boundary conditions has often been neglected. As a result, infinite plasma models, or models with static boundary conditions, are often applied to problems with variable boundary conditions. This gives completely erroneous results (examples are given in III).

In many theories, it is taken for granted that the behaviour of a plasma depends only on the local parameters (e.g., density, temperature, magnetic field). This can be quite misleading. As an example, in a non-curlfree (i.e., current carrying) plasma, the properties of the plasma are not *only* a function of the local parameters, but also of the outer *circuit* in which the current I closes (II.5). Figure II.16 shows a simple circuit consisting of an electromotive force V_b, a resistor R_0, and an inductance L. By changing R_0 and/or L, the behavior of the plasma may be changed in a drastic way. The value of R_0 decides whether the plasma is relatively stable or oscillating. If the plasma contains a double layer which explodes, the circuit energy $\frac{1}{2}LI^2$ is released in the layer. Hence, the violence of the plasma explosion is determined largely by the circuit.

The influence of the 'circuit' is essential, not only in a laboratory experiment, but also in space. In the latter case, the total volume in which the current flows affects the behavior of the plasma at every point. In many instances, it is convenient to introduce the boundary conditions by drawing an 'equivalent circuit' (II and III).

I.6. Cosmology and the Origin of the Solar System

In Chapters II, III and IV, the analysis of the properties of space plasmas is based on laboratory results and data obtained by *in situ* measurements in the magnetosphere (including the heliosphere). From such investigations, it is hoped to obtain an increasingly thorough knowledge of the general properties of space plasmas. It will never be possible to obtain an equally reliable knowledge of plasmas in extremely distant regions of space and time. However, as pointed out by Fälthammar *et al.* (1978), with the general plasma properties obtained from laboratory experiments and from *in situ* space measurements as a rather firm basis, it should be possible to approach the problems of the astrophysics of the more distant regions with much more confidence than with more or less speculative theories (as is often done today). Hence magnetospheric research is important not only to our knowledge of the magnetosphere, but also to our knowledge of phenomena in more distant regions.

In the last two chapters, some important astrophysical problems are approached by this method. In Chapter V, the origin and evolution of our solar system is summarized and in Chapter VI, the evolution and present structure of our universe (*cosmology*) is discussed.

I.7. Aims of the Monograph

The aims of the present monograph may be summarized as follows:

(1) Attention will be given to the question of how much knowledge can be gained by a systematic comparison of different regions of plasma. We will consider plasmas with linear dimensions which vary from laboratory size up to the Hubble distance (see Figure I.3).

(2) The traditional magnetic field description of plasmas will be supplemented by an electric current description (see Figure I.4). It is demonstrated that many problems are easier to understand with a dualistic approach.

(3) A rather strict distinction will be made between plasma regions which are accessible to *in situ* diagnostics and those regions which are not (see Figure I.3). In the former, we can expect to understand the plasma phenomena reasonably well, and to be able to discriminate between different interpretations, whereas in the latter regions, conclusions must necessarily remain largely speculative.

(4) It is claimed in this monograph that in the speculative domain, those theories, which are based on what we know from laboratory experiments and from regions accessible to spacecraft measurements, deserve higher credibility than those which are based on theories which we know to be misleading when applied to regions that have been subject to high-quality diagnostics.

As it is impossible to treat an extensive and complicated field like cosmic plasma physics in a rigorous way in a monograph of the present size, we have followed the policy that – as Kadomtsev (1965) puts it – "when rigour appears to conflict with simplicity, simplicity is given preference".

As it also is impossible to give a complete list of references to the enormous literature only a few typical results are quoted. More complete lists of references are usually found in the quoted papers.

CHAPTER II

ELECTRIC CURRENTS IN SPACE PLASMAS

II.1. Dualism in Physics

Since the beginning of this century, physics has been dualistic in the sense that some phenomena are described by a field formalism whereas others are treated in terms of particles. In the cosmic application of physics, we have frequent examples of this dualism: the propagation of waves is treated by Maxwell's equations, but the charge on a grain in space is derived by considering the photons and electrons which hit it. There are also phenomena, like the Doppler effect, which can be treated both from the wave and the particle point of view.

This is all very well understood, but it is not so well recognized how deeply this dualism penetrates into the field of cosmic plasmas. From Maxwell's first equation, which in a plasma can be written

$$\nabla \times \mathbf{B} = \mu_0 \mathbf{i}' \tag{1}$$

(where $\mathbf{i}' = \mathbf{i} + (\partial \mathbf{D}/\partial t)$ and in the future the prime will be dropped, compare Equation I(1)), we learn that we can describe electromagnetic phenomena in space either in terms of a magnetic field **B** or in terms of electric currents **i**. As magnetic fields are easy to measure, and, moreover, the mathematical treatment becomes simpler if i is eliminated, it seems obvious that we should use the field description and eliminate i. For example, when we treat waves propagating through a plasma, they are certainly associated with electric currents, but in many cases we can regard a current implicitly as the curl of the magnetic field and ignore it. However, in doing so we lose the particle aspect of the current; in other words, we neglect the fact that an electric current in space consists of motion of charged particles which have a certain mass, charge, and velocity, and which can often be considered as constituents of a gas with a certain temperature. Some of these properties can be formally introduced into the field description as bulk constants like the permittivity ϵ, the permeability μ and the conductivity σ, but this gives a poor and often misleading representation of the particle phenomena.

In contrast to this approach, the study of electric discharges, which starting a hundred years ago has clarified some essential properties of a plasma, approaches the phenomena from the particle aspect (motion of electrons and ions, formation of electrostatic double layers, establishment of non-Maxwellian velocity distribution, etc.). It is now obvious that several of the phenomena related to the particle aspect are of decisive importance in cosmic plasmas, and that by neglecting the particle aspect we deprive ourselves of the possibility of understanding some of the most important phenomena in cosmic plasma physics.

Further, the magnetic field description is often used in a careless way leading to the neglect of boundary conditions. Infinite models are applied to plasmas with finite dimension, resulting in erroneous conclusions. With the current description the whole *circuit* in which the current flows is included and in this way the neglect of the boundary conditions is more easily avoided.

In the following, we shall classify the plasma phenomena as magnetic field-related and particle-related (or electric current related), see Figure I.4. The former received much early attention, but the latter have only recently been brought into focus.

II.2. Particle-Related Phenomena in Plasma Physics

From the particle aspect, we can derive the equation of motion of a test particle by calculating the sum of all forces from other particles. In principle, we need not speak about fields at all. However, it is convenient to introduce the electric field E' by the equation

$$\mathbf{F} = \sum \mathbf{f} = e\mathbf{E}' \tag{2}$$

where $\mathbf{F}$ is the sum of all forces $\mathbf{f}$ acting on a test particle with charge e (neglecting gravitation and collisions).

The motion of a charged particle can be completely described as caused by the electric field $\mathbf{E}'$. A magnetic field exerts a negligible force on the particle at rest. However, if we make a relativistic transformation

$$\mathbf{E} = \mathbf{E}' - \mathbf{v} \times \mathbf{B} \tag{3}$$

from the coordinate system which moves with the particle velocity v in relation to a coordinate system at rest, we have in the latter coordinate system another electric field E. It is convenient to use a coordinate system at rest and describe the motion of the particle by the velocity v. In this coordinate system the force acting on the charged particle is

$$\mathbf{f} = e(\mathbf{E} + \mathbf{v} \times \mathbf{B}) \tag{4}$$

where $\mathbf{B}$ is given by Equation (1).

Hence, *seen from the coordinate system of the particle, B is unimportant*, but it is convenient to introduce this concept in order to make the calculation of $\mathbf{E}'$ easier. (This is a trivial statement but it is often forgotten.)

II.3. Magnetic Field Lines

A magnetic field line is by definition a line which is everywhere parallel to the magnetic field. If the current system changes, the shape of the magnetic field line changes, but *it is meaningless to speak about a translational movement of magnetic field lines.* The concept of 'frozen-in magnetic field lines' has played a central role in plasma physics due to the fact that in several situations, *but far from all*, it is legitimate to use it. The restrictions on its use are clarified in CE(5.4). Further, it has a quasi-pedagogical appeal. An impression is developed that you understand a situation even if in reality you have misunderstood it.

One of the requirements for using the 'frozen-in' field concept is that $E_{\parallel} = 0$. In order to satisfy this, the electric conductivity parallel to the magnetic field, $\sigma_{\parallel}$, must be infinite. If we use the classical formula (see, for example, CE, p. 149).

$$\sigma_{\|} = \gamma \frac{e^2 n_e \lambda_e}{m_e v_e} \tag{5}$$

we find that under cosmic conditions, $\sigma_{\|}$ is usually so large that we can regard it as infinite (e and m_e are the electronic charge and mass, n_e is the number density of electrons, λ_e and v_e are their mean free path and their thermal velocity, and γ is a constant of the order unity). However, this is not enough because there are a number of phenomena to which Equation (5) is not applicable. Basically these derive from the particle aspect of electric currents:

(1) Equation (5) is derived under the condition that the mean free path, λ_e, of electrons is small compared to the characteristic length of the variation of B, E, etc. This is often not satisfied in space plasmas. For example, in the outer magnetosphere and interplanetary space, it is not valid.

(2) When the electron drift velocity becomes equal to the sound velocity in the ionized plasma, a strong coupling may occur which transfers energy from the electrons to the sound waves; this causes an *anomalous resistivity*.

(3) If in a plasma the electric field is so strong that the electrons gain more energy than they lose by collisions, some electrons (*runaway electrons*) may be accelerated to very high velocities (Giovanelli, 1949; Dreicer, 1959, 1960; CE 4.3.4).

(4) If the velocity distribution in a plasma is non-Maxwellian, a magnetic field gradient may produce an electric field $E_{\|} \neq 0$.

(5) An electric current often produces electrostatic double layers (also called sheaths) associated with an almost discontinuous jump ΔV in the voltage (II.6).

As the aforementioned effects are common in low density cosmic plasmas (especially in 'collisionless plasmas'), the 'frozen-in' concept is very often invalidated. In particular, when combined with the 'magnetic field-line reconnection' concept ('magnetic merging'), it has led to a serious misunderstanding of many important phenomena (II.3.3, II.5.3 and CE 5.4).

II.3.1. MAGNETIC FIELD AND ELECTRIC CURRENT DESCRIPTION OF THE MAGNETOSPHERE

In order to demonstrate how unnecessary and misleading these concepts are, we shall treat the simple case of plasma flow in the magnetosphere, assuming a stationary state. This example will also be used as an illustration of the relationship between a magnetic field description and a particle description.

The stationary conditions of the magnetosphere are usually depicted by drawing the magnetic field lines (see Figure II.1). With the help of Equation (1), we translate this into a current and particle picture (Figure II.2). We place a current carrying coil in the Earth's interior which, if suitably designed, gives the Earth's magnetic field. If we are satisfied with the axisymmetric dipole component, then the coil may consist of a very small circular loop at the center of the Earth.

Similarly, we let the interplanetary magnetic field be produced by a current in a very large coil. In the simplified case of a south-directed interplanetary field, the coil may be a Helmholtz coil with its axis coinciding with the Earth's axis.

To these we must further add a number of coils representing the secondary magnetic fields. In a simplified model (Alfvén, 1977), these consist of a double coil representing

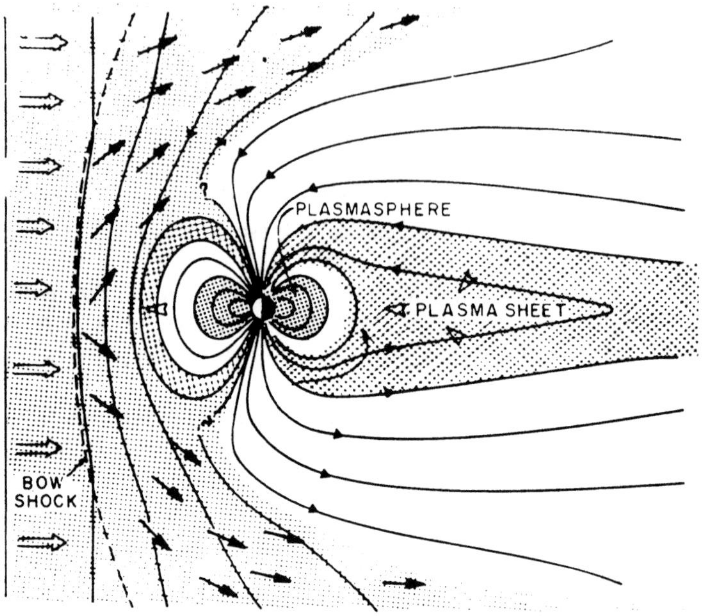

Fig. II.1. Standard model of solar wind and magnetosphere (after Wolf, 1975).

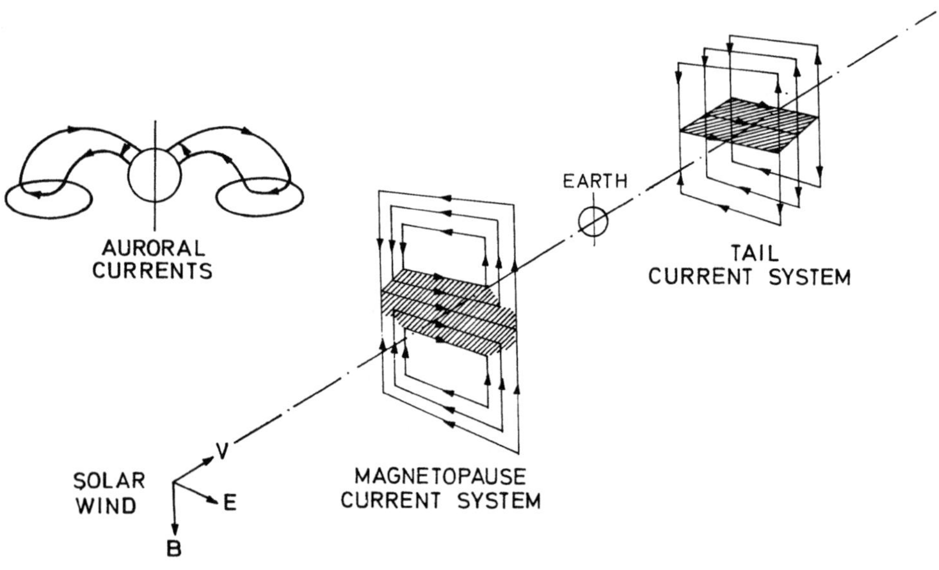

Fig. II.2. Simplified current system which gives the magnetic field of Figure II.1. The auroı currents are enlarged and depicted separately.

the magnetopause current system, a double coil representing the tail current system, and a system of coils representing the ionosphere–magnetosphere current system (see Figure II.2, where the latter is depicted separately). By making the coils of very thin electrically insulated metal wires, we can obtain a good approximation to the current system which in reality consists of distributed currents (a realistic current system is derived in Section III.6).

Similarly, we produce the electrostatic field by a number of charges placed in suitably fixed positions. For example, in the simple case $v =$ const., $B =$ const, the interplanetary electric field as seen from an earth-centered system is $\mathbf{E} = -\mathbf{v} \times \mathbf{B}$. This electric field can be produced by two condenser plates separated by a large distance d and charged to a potential $V = d \cdot E$. Other electric fields created by space charge should be represented in a similar way.

In the stationary state we have considered above, *both the electric and the magnetic fields are static*. We can depict the magnetic field by drawing the magnetic field lines (see again Figure II.1), but it should be observed that a magnetic field line has the Maxwellian meaning. It is a line which everywhere has the direction of the magnetic field. *To ask whether a field line 'moves' or not makes no sense*. In our static vacuum model, it is natural to depict them as immovable in relation to the coils which produce them, which means in relation to the earth.

II.3.2. PARTICLE MOTION IN THE MAGNETOSPHERE

So far our model does not contain any movable charged particles (outside the wires). In this vacuum model, we inject one charged test particle, either in interplanetary space or in the ionosphere. Its motion is completely determined by the electric and magnetic fields. As the magnetic field is static, the energy W of the particle is given by

$$W = W_0 + e \int \mathbf{E} \cdot \mathrm{d}\mathbf{s}, \tag{6}$$

where W_0 is the initial energy, ds the line element, $\mathbf{E}$ the (static) electric field from the fixed charges, and e the charge of the particle.

Next, we inject a large number of solar wind particles (and particles from the ionosphere), but still only a negligible fraction, ϵ, of what corresponds to the real case. Assuming that the mutual collisions (as well as the collisions with the model structure) are negligible, they will behave as a number of test particles. If our model is designed correctly, they will increase the space charge given by the fixed charges of the model by the fraction, ϵ, and their flow close to the coil wires will increase the magnetic field by the same fraction. If we reduce all coil currents and all fixed charges by ϵ, we will return to the same electric and magnetic field as before the injection. Hence, in this simple way, the electromagnetic fields in a stationary magnetosphere are described exclusively by electric currents and electric charges. A model based on this principle is worked out in III.6.

We now slowly increase ϵ to unity. At the same time we reduce the coil currents and the fixed charges so that eventually they become zero. It is easily seen that this can be done in a way that allows the electric and magnetic fields to remain constant. We can now remove the model structure, and every particle will still move and change its energy as if it were a single test particle in the vacuum model. Our model now depicts how plasma in our surroundings flows and changes its energy.

II.3.3. CONCLUSIONS ABOUT 'FIELD LINE RECONNECTION' AND 'MERGING' IN THE STATIONARY MAGNETOSPHERE

Our Gedanken experiment shows that neither the injection of one test particle, a small number of test particles, or all of the solar wind particles call for a change in the Maxwellian concept of magnetic field lines. There is no need for 'frozen-in' field lines moving with the plasma, still less for 'field-line reconnection' or 'magnetic merging'. The magnetic field always remains static and not a single field line is 'disconnected' or 'reconnected'. The energy of a charged particle is given by Equation (6). There is no 'field-line reconnection' that can transfer energy to the particles or release energy in any other way. Other arguments against reconnection models are forewarded by Heikkila (1978).

If the magnetic field varies with time, the geometry near the neutral points (points where $B = 0$) may change in a way that may be considered as the field lines disconnecting and reconnecting. It may be argued that in this case, the usual field-line reconnection formalism should be applicable. As will be shown in II.5 this is not correct. The field-line reconnection theories are erroneous also in this case.

II.4. Filaments

II.4.1. OBSERVATIONS OF FILAMENTS

Filamentary structures are often observed in cosmic plasmas. There seems to be a continuous transition from true filamentary structures to sheet structures, via structures of elliptical cross-section. Consequently, from a phenomenological point of view a clear distinction between filaments and sheets is not called for.

In the region accessible to *in situ* measurements, there are the following filamentary structures, all of which are observed to be associated with, or are likely to be associated with electric currents:

(a) In the aurora, *filaments* parallel to the magnetic field are very often observed ('auroral rays' – see Figure II.3). They are sometimes very thin, with thickness down to ~ 100 m. They occur under conditions that suggest that they are due to *Birkeland currents*. However, *in situ* measurements have not yet completely clarified the relation between the visual structures and the electric currents. This also holds for the often very thin *auroral arcs* and *draperies*. The *auroral electrojet* itself is of a filamentary character, and of course, there is no doubt that it carries a current.

(b) *Inverted V events* and the *in situ* measurements of strong electric fields in the magnetosphere (II.7), especially at heights of one Earth radius above the surface, demonstrate the existence of filamentary structures. These structures, which often have elliptical or sheet-like cross-sections are likely to be produced by Birkeland currents.

(c) In the *ionosphere of Venus*, 'flux ropes' are observed with a structure which shows beyond any doubt that they are produced by filamentary currents. Their diameters are typically 20 km (see II.4.7).

In regions which are not accessible to *in situ* measurements, we often observe filamentary structures:

(d) In *the Sun*. Prominences, spicules, coronal streamers, polar plumes, etc. (see Figures II.4 and II.5). In all these cases there are more or less convincing arguments for attributing the filaments to field-aligned currents.

Fig. II.3. Auroral rays in a corona. This pencil sketch by Nansen (in December 1894) gives a better picture of the filamentary structure than photographs, because of the long exposure time needed compared with the fluctuations in the auroral rays.

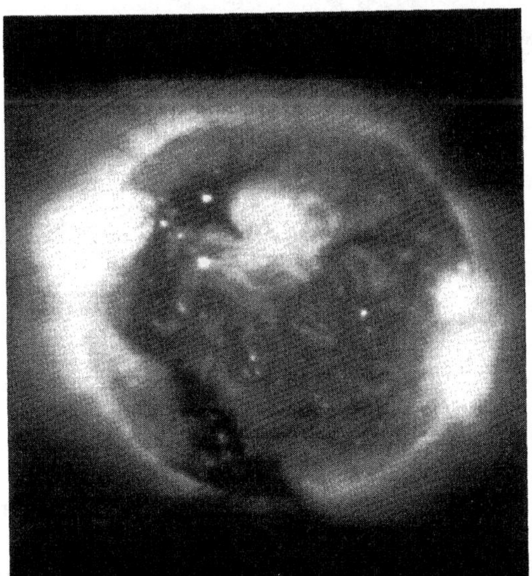

Fig. II.4. Soft X-ray photograph of the Sun from Skylab (1973). A number of loops are seen on the disc which are identified with high density regions. The looplike structures may be caused by electric currents. (American Science and Engineering.)

Fig. II.5. Solar corona (March 1970). The filamentary structure is probably caused by electric currents. (G. Newkirk, High Altitude Observatory.)

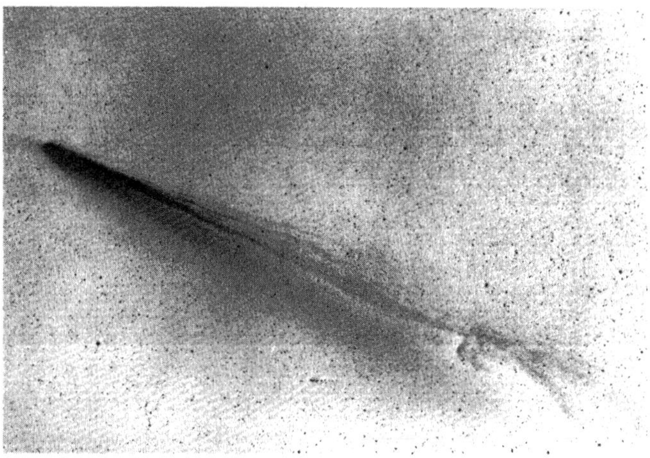

Fig. II.6. Comet Kohoutek (Jan 1974). The filamentary structure of the tail of a comet is according to Mendis (1978), due to electric currents.

Of special interest is the fact that the application of the contrast enhancement method displays the filamentary structure still more clearly. (Of course, one should be aware of the fact that this technique may produce spurious structures.)

(e) *Cometary tails* often have a pronounced filamentary structure. There is good reason to interpret them as due to field-aligned currents (Mendis, 1978). (See Figure II.6.)

(f) In the *interstellar medium* there is an abundance of filamentary structures (see Figures II.7 and II.8). Some of these are attributed to shock the front sent out by nova or

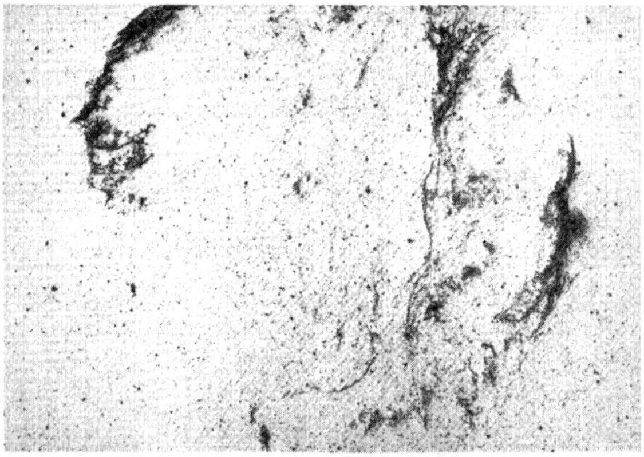

Fig. II.7. The Veil nebula in Cygnus. Its filamentary structure is probably caused by a supernova explosion.

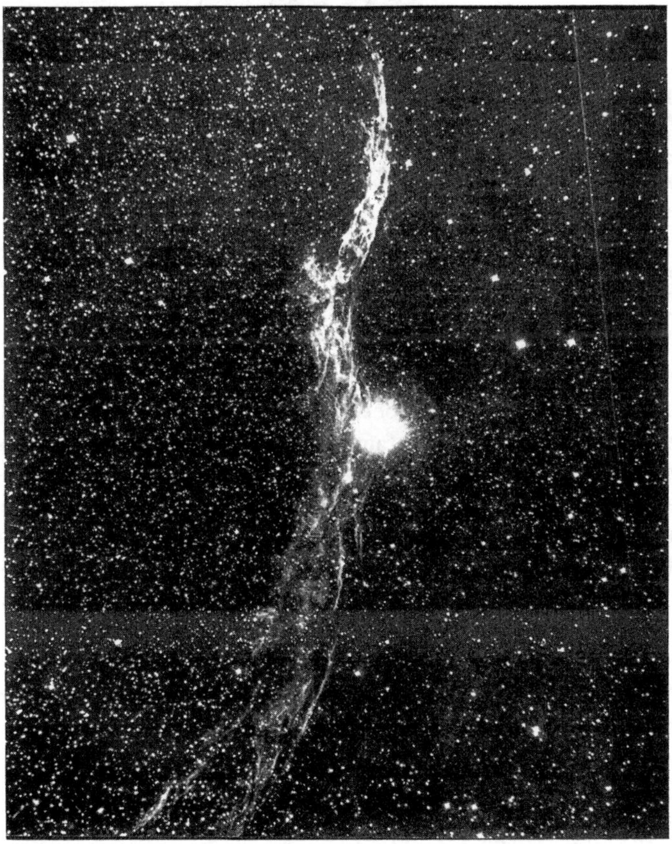

Fig. II.8. Detail of the Veil nebula (*Sky and Telescope*, November, 1979).

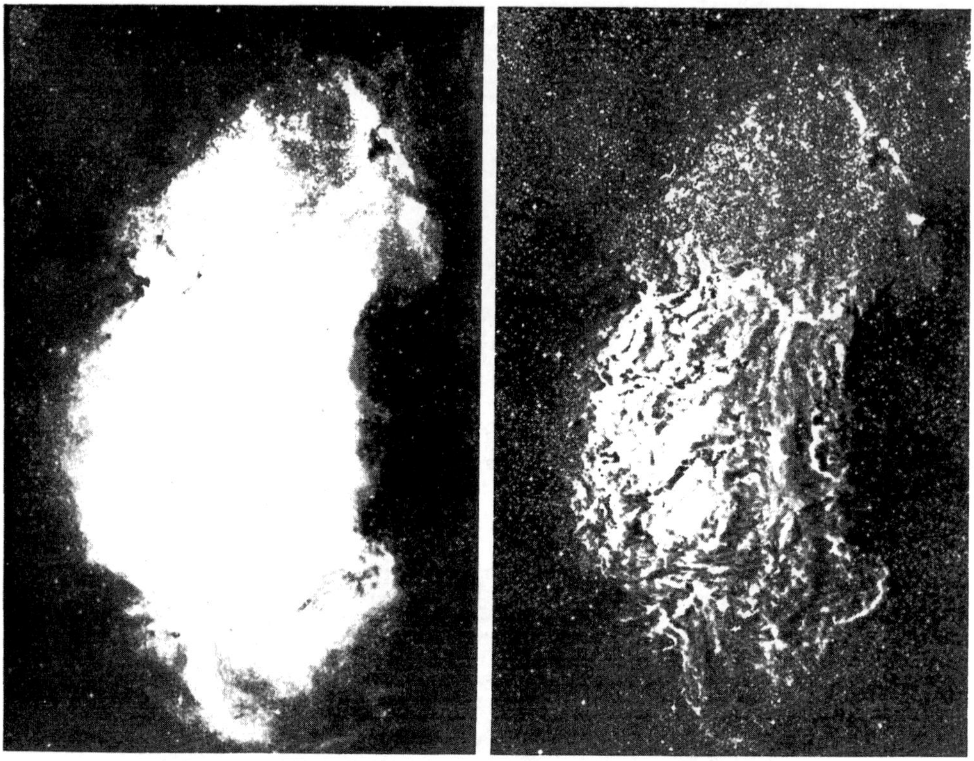

Fig. II.9. Filamentary structures are much more common than usually thought as shown by comparing the ordinary photograph of the Lagoon nebula in Sagittarius (left) with the same photograph subjected to contrast enhancement (right) (*Sky and Telescope*, April, 1979).

supernova explosions. As a hydromagnetic shock wave is necessarily associated with electric currents, there is no conflict between the shock wave and the electric current interpretations. On the contrary, the fact that we observe filaments where we expect to find current-carrying shock fronts, speaks in favor of the electric current interpretation.

(g) *Interstellar clouds* (or nebulae) which look smooth and rather homogeneous in ordinary pictures, often reveal a filamentary structure when subjected to *contrast enhancement* (see Figures II.9 and II.10). With the current interpretation of filamentary structure, this may be regarded as an indication that they consist of a network of currents (which from other considerations is quite likely (see V.3)).

In addition to the electric current interpretation of filaments, we must be aware of alternative interpretations. It has been suggested that the filaments are due to hydrodynamic shear motion. For example, if a drop of ink is injected in a glass of water in which the water is in turbulent motion, the drop is often dragged out into thin filaments. Another example of filamentary structure is cirrus clouds.

These examples refer to hydrodynamic motions in regions where electromagnetic forces are of no importance. However, in the entire region subject to *in situ* measurement and dominated by plasma effects, there does not seem to be a single example of a similar

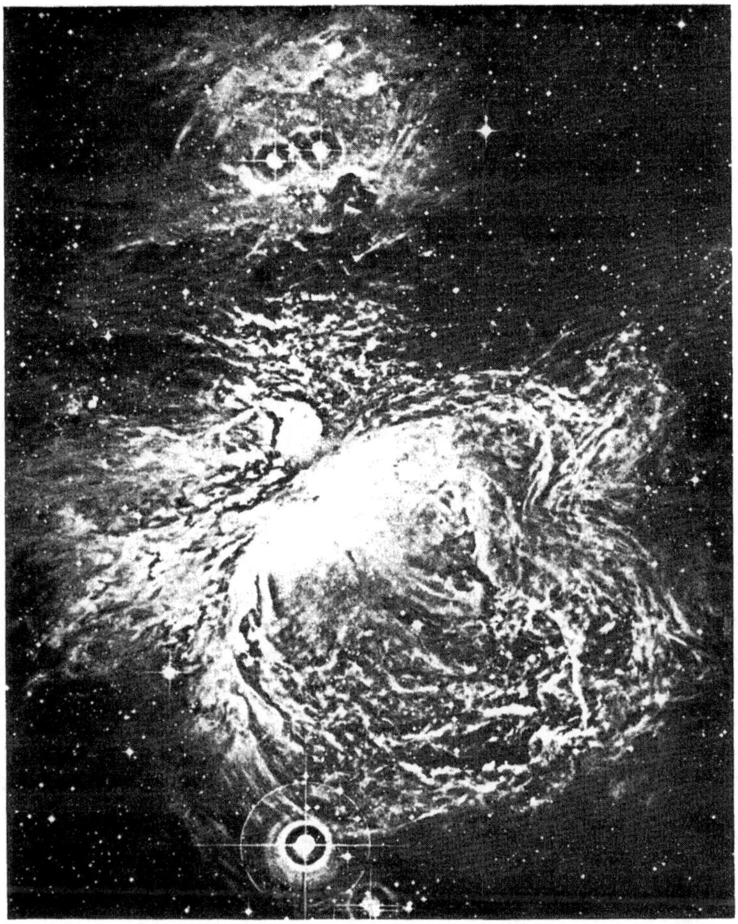

Fig. II.10. Contrast-enhanced photograph of the Orion nebula (*Sky and Telescope*, April, 1979).

effect. None of the observed filaments can be regarded with any certainty as resulting from such effects, whereas in all cases there is either convincing proof, or reasonably strong evidence for an electric current interpretation. Outside of the region accessible to spacecraft, we cannot decide with any confidence which of the interpretations is correct. However, if we follow our general approach, *viz.* to see how far we can explain phenomena in more distant regions by means of mechanisms which we know are working in the better explored regions, we should give the current interpretation the highest priority.

II.4.2. CONSTRICTION OF A DISCHARGE

A discharge may fill all of the space between the electrodes or it may be confined to a narrow channel. The former is the case for a laboratory glow discharge at low pressure. Examples of the latter case are arcs, sparks, flashes of lightning, and the 'pinched discharges' studied in thermonuclear research.

In laboratory discharges, especially at low ionization, the constriction is often connected with a 'falling characteristic', by which we mean that the electric field necessary to maintain the discharge is a decreasing function of the current density (II.5). If the total discharge current is constant, the electric field strength is smaller when the current is concentrated in a small channel than when it fills the whole space. A discharge often adjusts itself so that the electric field strength becomes a minimum.

The constriction of a discharge is a phenomenon which is influenced by several different factors (see von Engel and Steenbeck (1932, 1934), which still probably is the best introduction to this field).

A *thermal constriction mechanism* is operative in arcs.

An arc plasma obeys rather complicated laws, the essence of which is that the heat produced by the electric current shall cover the thermal losses, mainly through convection, conduction and radiation to the colder surroundings. It can be shown that as a result, discharges are usually constricted at atmospheric pressure, whereas at a pressure below, say, 1 mm Hg, constriction becomes a rarer phenomenon.

The thermal constriction mechanism is not primarily dependent on a magnetic field. However, *magnetic effects* may give rise to constrictions of different types.

Currents perpendicular to B may be constricted. Examples of such phenomena in the ionosphere are the *equatorial electrojet and the auroral electrojet*.

However, in cosmic plasmas the perhaps most important constriction mechanism is the *electromagnetic attraction* between parallel currents. A manifestation of this mechanism is the *pinch effect*, which was studied by Bennett long ago (1934), and has received much attention in connection with thermonuclear research. As we shall see, phenomena of this general type also exist on a cosmic scale, and lead to a bunching of currents and magnetic fields to filaments or 'magnetic ropes'. This bunching is usually accompanied by an accumulation of matter, and it may explain the observational fact that cosmic matter exhibits an abundance of *filamentary structures* (II.4.1). This same mechanism may also evacuate the regions near the rope and produce regions of exceptionally low densities (II.4.6).

The electromagnetic attraction including the pinch effect has been very thoroughly studied in the laboratory. Indeed, the magnetic mirror technique used for thermonuclear plasma confinement is based on the pinch effect. In cosmic physics, however, it is very often forgotten. For example, the formation and evolution of interstellar clouds is normally treated by neglecting this effect, in spite of the fact that it must necessarily be of decisive importance (IV.8 and V.4). This has led to the regrettable result that this field and quite a few other fields of astrophysics have run into dead ends.

II.4.3. PINCH EFFECT. THE BENNETT RELATION

Consider a fully ionized cylindrical plasma column with radius r, in an axial electric field E, which produces an axial current of density i, Figure II.11 (CE 5.5.2). The axial current is associated with an azimuthal magnetic field. The current flowing across its own magnetic field exerts a force $\mathbf{i} \times \mathbf{B}$, which is directed radially inward, and causes the plasma to be compressed towards the axis (hence the name 'pinch effect'). For the equilibrium between the compressing electromagnetic force and the sum p of the electron and ion pressures, p_e and p_i, we have

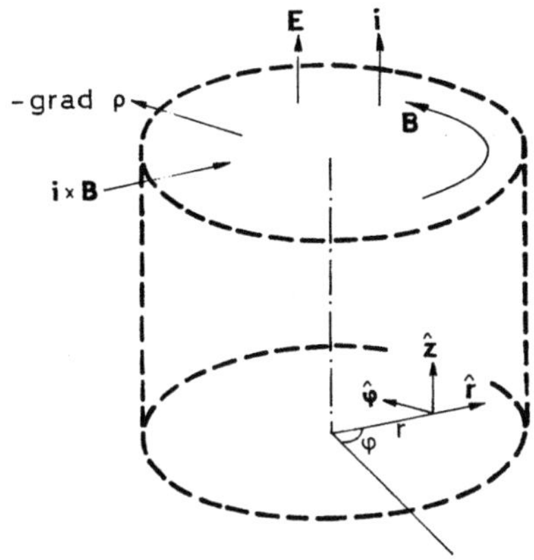

Fig. II.11. In the stationary pinched discharge the electromagnetic force $\mathbf{i} \times \mathbf{B}$ (attraction between parallel currents) is compensated by the gradient of the plasma pressure.

$$\nabla p = \nabla(p_e + p_i) = \mathbf{i} \times \mathbf{B}. \tag{7}$$

The axial electric field is perpendicular to the azimuthal magnetic field produced by the axial current. In spite of this the current density is essentially given by

$$i = \sigma E \tag{8}$$

as soon as a stationary state has been reached. This is true because in the stationary state no radial Hall current can flow.

By employing the Maxwell equation $\nabla \times \mathbf{B} = \mu_0 \mathbf{i}$, Equation (7), and the perfect gas law, we quickly arrive at the *Bennett relation* (for details, see IV.8)

$$2Nk(T_e + T_i) = \frac{\mu_0}{4\pi} I^2 \tag{9}$$

where N is the number of electrons *per unit length* of the column, T_i and T_e, the ion and electron temperatures, I, the total current in the column, and k Boltzmann's constant.

II.4.4. FILAMENTARY CURRENTS IN FORCE-FREE MAGNETIC FIELDS

In a laboratory pinched discharge the electric current can flow across its own magnetic field because the electromagnetic forces are balanced by the plasma pressure, and the radial motion vanishes. In *cosmic plasmas* the pressure is often negligible and hence the magnetic field force-free.

Consider a medium-density plasma with the configuration shown in Figure II.12. The electric field is homogeneous and directed along the z-axis. The magnetic field, which derives partly from currents in the plasma and partly from external sources, has both ϕ- and z-components.

Under the influence of the electric and magnetic fields, both electrons and ions drift with the velocity

$$\mathbf{v} = (\mathbf{E} \times \mathbf{B})/B^2 \tag{10}$$

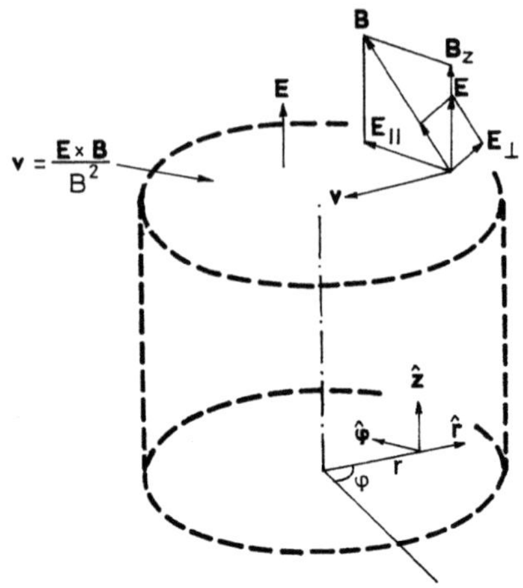

Fig. II.12. When the pressure is negligible the plasma acquires a drift velocity **v** such that the electric field in the moving plasma is parallel to **B**. Therefore current flows only along the magnetic lines of force.

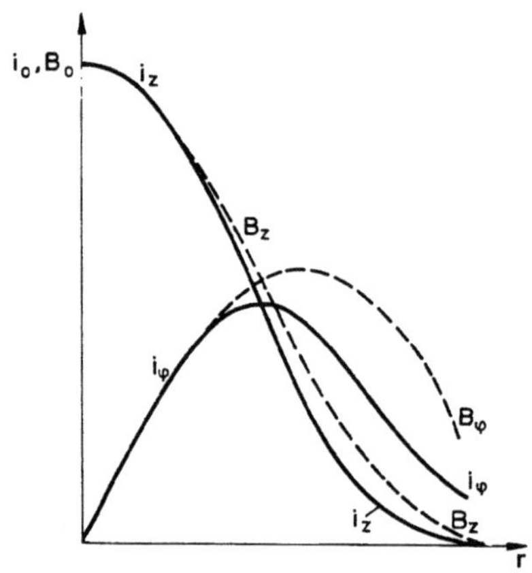

Fig. II.13. The current density components are i_ϕ and i_z, the magnetic field components B_ϕ and B_z. i_0 is the current density amplitude and B_0 the magnetic field strength, both along the z-axis.

so that the plasma as a whole moves radially inward. If we transform to the plasma frame, we find that the electric field is parallel to the magnetic field, and consequently, current only flows along the magnetic field lines. Moreover, since the magnetic force is given by $\mathbf{i} \times \mathbf{B}$, it vanishes, and the magnetic field is force free.

It is shown in CE (5.5.3), that the current in this configuration is constricted to a cylinder and forms a line current. Also shown is that the axial magnetic field is similarly constricted, and as a result the field at large distance is azimuthal (see Figure II.13). Indeed, the qualitative field line pattern, as is shown in Figure II.14, suggests the name 'magnetic rope'.

In the present treatment we have assumed a constant conductivity, but Murty (1961) has shown that similar results hold for force-free fields under much more general conditions.

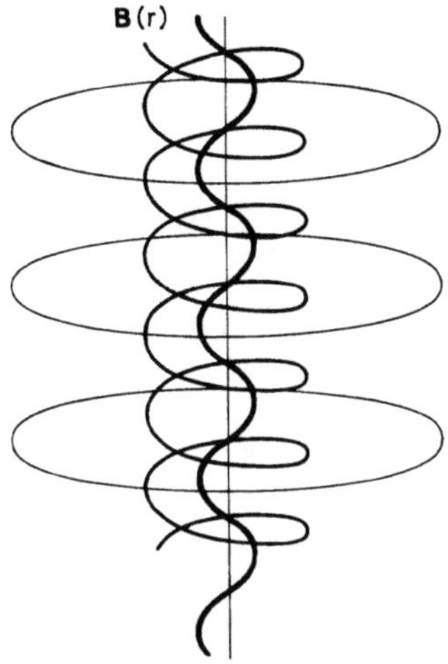

Fig. II.14. Magnetic field lines at three different radii in a magnetic rope.

As mentioned above, the electric and magnetic fields cause an inward drift motion by which matter is accumulated. 'Magnetic ropes' should therefore tend to coincide with material filaments that have a higher density than the surroundings. The cosmic 'magnetic ropes' are not observable themselves, but the associated filaments of condensed matter can be observed by the radiation they emit and absorb. The electric current in a magnetic rope dissipates energy so that we should expect a rope to be hotter than the surroundings. However, this conclusion is not necessarily valid, because the increased density may result in radiative cooling which may make the rope cooler than the surroundings. An example of this is the prominences, which are much cooler than the surrounding corona because their density and hence their radiation losses are much larger than that of the surrounding medium. The constriction mechanism discussed here may therefore explain the filamentary structures that are observed to occur so abundantly in cosmic plasma (filamentary interstellar clouds, coronal filaments, prominences, etc.), as demonstrated in II.4.1. As pointed out in II.4.2, the neglect of electromagnetic attraction has caused quite a few fields of astrophysics to waste effort in approaches which are basically wrong.

II.4.5. THEORY OF MAGNETIC ROPES

Marklund (1978) has analyzed the structure of filamentary currents, 'magnetic ropes', collecting ionized gas from the surroundings (Fig. II.15). He has calculated the resulting stationary state when the inward drift of ions and electrons towards the axis of a filament is matched by recombination and outward diffusion of the neutral gas. The equilibrium density of the ionized component normally has a maximum at the axis. However, under certain conditions the filament may be a cylindrical shell with an ion density minimum at the axis. Magnetic ropes have been observed by *in situ* measurements in the ionosphere of Venus (see III.5.3).

In the case of a partially ionized gas mixture, a temperature gradient will cause the radial transport to depend on the ionization potential, so that in general, the elements with the lowest ionization potential are brought closest to the axis. We may expect the

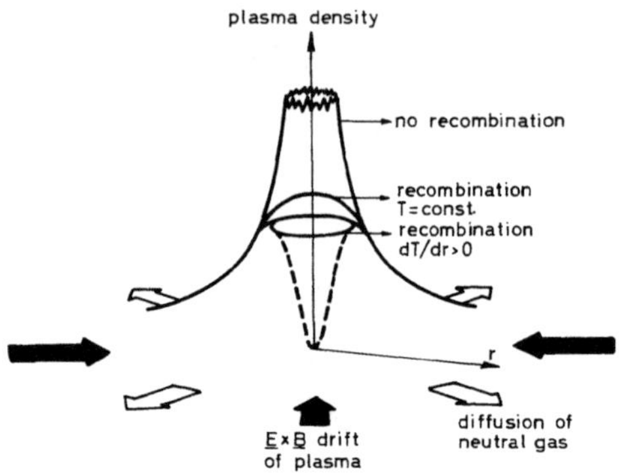

Fig. II.15. Number density profile as a function of radius sketched qualitatively (in three dimensions) for three cases: no recombination, recombination with $T =$ const., and recombination with lower central temperature (Marklund, 1978).

elements to form concentric hollow cylinders whose radii increase with ionization potential.

Quite generally, it seems likely that for a rather wide range of parameters, a current through a partially ionized plasma is able to produce element separation (IV.3).

II.4.6. ION PUMPS

The drift of ionized matter from the surroundings into the rope means that the rope acts as an *ion pump*, which evacuates the surroundings. Regions with extremely low densities can be produced in this way. It is possible that 'coronal holes' are produced by this mechanism.

II.4.7. FLUX ROPES IN THE IONOSPHERE OF VENUS

Elphic *et al.* (1979) have made *in situ* measurements of the current system in the ionosphere of Venus. They observe a 'flux rope structure' which seems to be similar to the flux ropes described in II.4.5. During quiescent conditions, the flux ropes are observed to be a nearly ubiquitous feature of the dayside ionosphere. Their diameters are of the order 15–20 km. The data indicate only a small thermal pressure gradient, so that $\mathbf{v} \times \mathbf{B}$ should be small. The observed magnetic field structure can be compared with the theoretical field structure of a flux tube. It is evident that the agreement is satisfactory.

This is the first convincing observational evidence for the theoretically expected structure of filamentary currents producing flux ropes. Due to its basic importance for the electric current structure of cosmic plasmas, further investigations of the flux ropes in the ionosphere of Venus are very important. One should also try to identify similar structures in other regions.

II.5. Local Plasma Properties and the Circuit

Consider a plasma tube, in which the current I, originating from a battery with e.m.f. V_b, is transmitted by a circuit with resistance R_0 and inductance L (Figure II.16). The voltage $V(I)$ between the electrodes is a function of I and of plasma parameters like

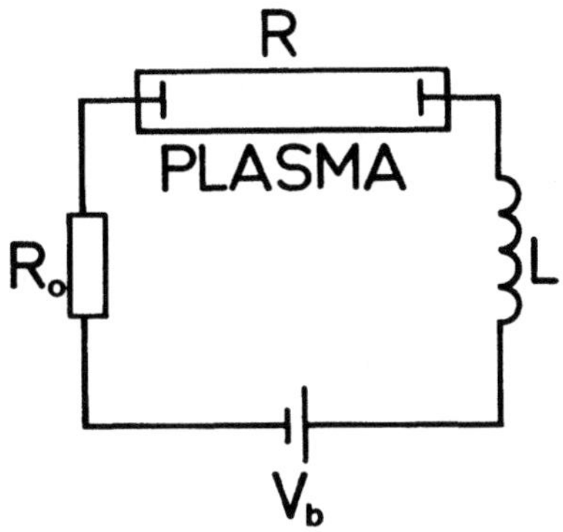

Fig. II.16. A plasma tube in a circuit. The behavior of the plasma depends on the circuit. If the plasma has a negative resistance R, it will oscillate if $R_0 + R < 0$. Hence by varying R_0 we can control the oscillations. Further, the inductance L determines this frequency and the total energy release should the current through the plasma be disrupted. Hence *even if we know all the plasma parameters* (temperature, pressure, magnetisation etc.) *in the tube*, we cannot predict the behavior of the plasma unless we know the parameters of the circuit of which the plasma is a part.

density, magnetic field, temperature, etc., which depend on I in a complicated way. If, to the first approximation, we set $V(I) = V_0 + R(I - I_0)$, where I_0 and R are constants, and V_0 the voltage between the electrodes at $I = I_0$. The circuit then obeys the equation

$$L \frac{dI}{dt} = V_b - V_0 - (R + R_0)(I - I_0). \tag{11}$$

If $R + R_0 = 0$, the plasma will always be in equilibrium $V_0 = V_b$. If $R + R_0 > 0$ and $V_0 = V_b$, the current will decrease until an equilibrium is reached. However, if $R + R_0 < 0$ (which it often is, because R is frequently negative), an equilibrium is impossible. As a result, the plasma may produce regular oscillations with the frequency f, or the current may go to zero with a certain time constant. In the latter case, the discharge may either be automatically reexcited or left extinct (see II.6.1 and II.6.5).

Hence, *the behavior of the plasma depends on the outer circuit*. A stable plasma discharge may be made unstable by decreasing R_0. The frequency of the oscillations can be regulated by changing R_0 or L, and of course, V_b is of decisive importance.

In the case of the instability leading to an extinction of the current, it should be remembered that *every electric circuit is explosive* in the sense that if we try to disrupt the current, a release of the whole inductive energy

$$W_L = \tfrac{1}{2} L I^2 \tag{12}$$

at the point of disruption will occur. This is a well-known phenomenon in high-power transmission lines; switching off the current necessarily leads to an explosion which must be absorbed by the switch.

If the current disruption is caused by an instability in the plasma, *the inductive energy W_L in the circuit will be released in the plasma*. It then causes an explosion which may be violent if W_L is large. As we shall see, the disruption of a current through a plasma is often caused by a double layer becoming unstable.

II.5.1. BOUNDARY CONDITIONS

Consider a volume V of space limited by a surface S. The properties of the plasma inside of S depend on the boundary conditions. Thus by changing the current through S, we can change the behavior of the plasma. As was seen in the preceding paragraph, the properties of a plasma depend on the whole circuit in which the current flows. This means that we can describe the plasma inside of S by parameters inside S only if $i = 0$ everywhere on S.

Hence, even if we know all plasma parameters (like density, temperature, and magnetization) at every point inside S, we can describe the plasma properties theoretically only if *there is no electric current crossing the surface*. Therefore, the boundary problems have to be analysed very carefully.

II.5.2. 'INVISIBLE' TRANSFER OF ENERGY

The electric current description shows that it is possible to *transfer energy in an 'invisible' way*. As an example we can take the transfer of energy in the auroral circuit (see Figure II.17) which transfers energy from a moving plasma cloud C to the region of release D through electric currents in B_1 and B_2; an electric voltage difference exists between B_1 and B_2 (see further III.2). Both the current and the voltage difference are difficult to detect even by *in situ* measurements. From a distance, it is almost impossible to observe them. What we may observe from a distance is that energy disappears in a region C (e.g., the velocity of the cloud is decelerated) and that energy is released in another region D (e.g., by the emission of radiation from this region). Hence, we have a mechanism

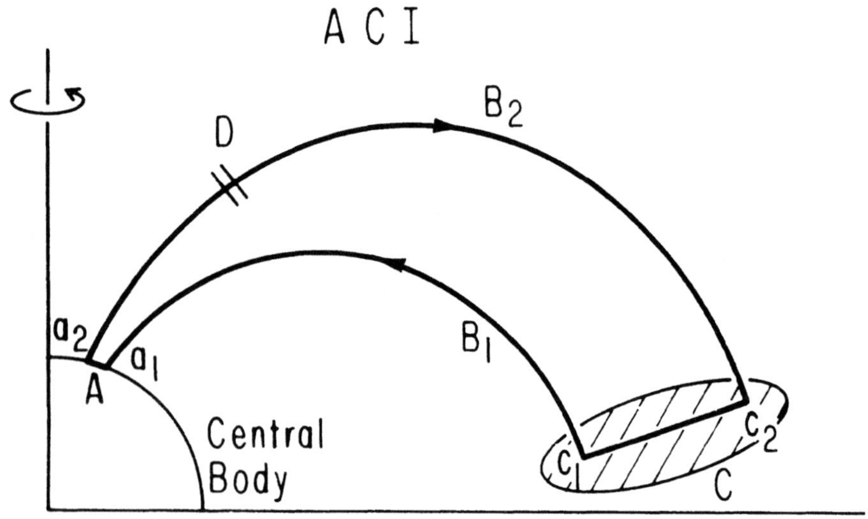

Fig. II.17. Auroral circuit (seen from the Sun). The central body (Earth and ionosphere) maintains a dipole field. B_1 and B_2 are magnetic field lines from the body. C is a plasma cloud near the equatorial plane moving in the sunward direction (out of the figure) producing an electromotive force $V = \int_{C_1}^{C_2} \mathbf{v} \times \mathbf{B} \cdot d\mathbf{s}$ which gives rise to a current in the circuit C1, a1, a2, C2 and C1. In a double layer D with the voltage V_D, the current releases energy at the rate $P = IV_D$, which essentially is used for accelerating auroral electrons. The energy is transferred from C to D not by high energy particles or waves, and *not* by magnetic merging or field reconnection. It is a property of the *electric circuit* (and can also be described by the Poynting vector).

of 'invisible' transfer of energy over a large distance. A possible example of this is the energy release in double radio sources (III.4.4).

An 'invisible' transfer is a characteristic of many electric currents. For example, when we observe the Earth from a satellite or an aircraft, we easily see street lights dissipating electric energy, and we may also see hydroelectric power stations generating it, but it is almost impossible to observe how the energy is transmitted. In important respects, the transfer of energy in cosmic physics is very similar to the electrotechnical transfer.

II.5.3. 'MAGNETIC MERGING' THEORIES

What we have found means that we can describe plasma phenomena inside a finite volume *only if no electric current crosses the surface*. In the terminology of the magnetic field description, this means that we can describe plasma phenomena inside a finite volume only if *the perpendicular component of the curl is zero at every point of the surface.*

All theories of 'magnetic merging' (or 'field line reconnection') which do not satisfy this criterion are misleading or erroneous, and deserve no attention. This does not mean that *all* papers in which 'magnetic merging' is used are of no interest, because there exist some good papers (e.g., Hill, 1975) in which the term is merely a synonym for "current-sheet acceleration."

II.6. Electric Double Layers

II.6.1. GENERAL PROPERTIES OF DOUBLE LAYERS

In a low density plasma, localized space charge regions may build up large potential drops over distances of the order of some tens of the Debye lengths. Such regions have been called *electric double layers*. An electric double layer is the simplest space charge distribution that gives a potential drop in the layer and a vanishing electric field on each side of the layer (Figure II.19). In the laboratory, double layers have been studied for half a century, but their importance in cosmic plasmas has not been generally recognized until recently. A number of investigations have confirmed the existence of double layers under various conditions, both in the laboratory and in the magnetosphere (Coakley *et al.*, 1978; de Groot *et al.*, 1977; Hubbard and Joyce, 1979; Leung *et al.*, 1979; Quon and Wong, 1976). Recent reviews of double layer phenomena of importance to cosmic plasmas include two general surveys (Block, 1978; Fälthammar, 1979) as well as separate reviews of experiments (Torvén, 1979, 1980) and theory (Carlqvist, 1979a).

Double layers may be produced in a number of ways. In general, a plasma screens itself from walls and electrodes by producing double layers, which in this case often are referred to as *sheaths*. In addition, if the electron temperature, for example, is different in two regions of the plasma, the transition between them is often not smooth. Instead, the plasma divides itself into two (or several) homogeneous regions, separated by one (or several) double layer.

At low current densities, the thermal motion of the electrons is superimposed on a relatively small drift motion. If the current density increases so that the drift velocity equals the thermal velocity, a double layer is usually produced. This is due to an instability: the electron motion assumes the character of a beam, and the result is a phenomenon related to the two-stream instability.

Double layers in nonmagnetized plasmas have been extensively studied in the laboratory. Because of the possible effect of the walls in laboratory experiments, it has been questioned whether these studies are applicable to cosmic situations. In recent experiments (Lutsenko *et al.*, 1975; Coakley *et al.*, 1979; Torvén and Andersson, 1979), double layers have also been studied in magnetized plasmas. The plasma is kept away from the walls by a magnetic field so that wall effects are of no importance. It is found that far away from the walls, double layers are formed in these magnetized plasmas as well. This is essential for the application of the laboratory results to cosmic plasmas. However, most of the laboratory investigations have been made with magnetic fields which make the electron Larmor radius, but not the ion Larmor radius small compared to the size of the vessel. As in cosmic plasma problems also the ion Larmor radius often is small compared to the characteristic length, the validity of the cosmic application is still not completely clear.

A double layer is usually *noisy*; in other words, it produces oscillations within a large frequency band. Often the amplitude of the noise is small compared to the voltage drop across the double layer, and this can be considered as (almost) static. A detailed investigation of such double layers has recently been made by Baker *et al.* (1980). However, a double layer may also become unstable and *explode*, by which we mean that the voltage drop suddenly increases by orders of magnitude. A possible mechanism is given by Carlqvist (1969). The character of the explosion is largely determined by the inductance L and the resistance R of the circuit in which the current flows. (Exploding double layers were first discovered and studied in mercury rectifiers used in d.c. high-power transmission circuits.) The phenomenon is likely to be basic for the understanding of solar flares, magnetic substorms and related phenomena.

At the explosion, the layer voltage is mainly given by $-L\, \mathrm{d}I/\mathrm{d}t$, which may exceed the normal double layer voltage V_D by several orders of magnitude. The process usually continues until I vanishes and the energy $\frac{1}{2}LI_0^2$ (I_0 is the initial current) is released in the double layer, mainly in the form of kinetic energy of ions and electrons. The electrons and ions later produce heat and radiation as they interact with the matter in their neighborhood.

Both electrons and ions are accelerated in double layers. For strong double layers in cold plasmas, the ratio between the ion current i_i and the electron current i_e is expected to be

$$\frac{i_i}{i_e} = \left(\frac{m_e}{m_i}\right)^{1/2}. \tag{13}$$

However, in many cases this relation is not valid (Block, 1972).

Measurements of a double layer in a magnetized laboratory plasma (Coakley *et al.*, 1979; Torvén and Lindberg, 1980) show that the equipotential surfaces, which are perpendicular to the magnetic field in the layer, curve at the plasma boundary, where they become parallel to the magnetic field. In this way, the plasmas on each side of the layer become surrounded by regions with strong radial electric fields perpendicular to the magnetic field. This radial field is directed inward on the low potential side of the layer and outward on the high potential side. The configuration resembles two 'cables' connected by a double layer.

As an example of double layer investigations in the laboratory, Figure II.18 shows the experiments of Torvén and Lindberg (1980). Figure II.19 shows the potential profile

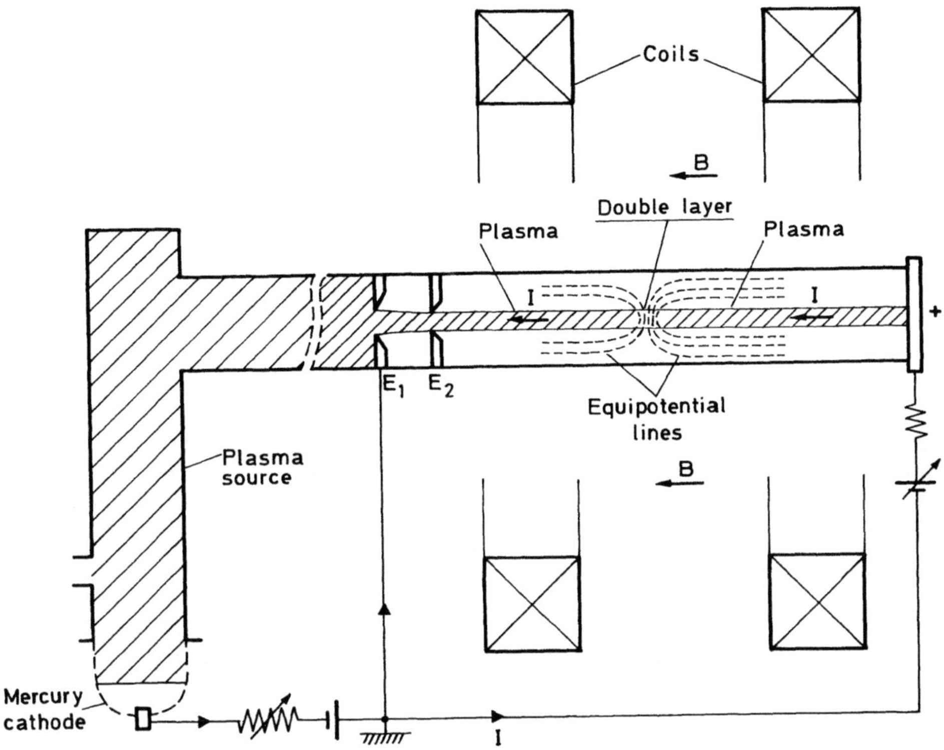

Fig. II.18. From a mercury plasma produced by a discharge, a plasma column confined by a magnetic field penetrates to the anode. In this current-carrying column a double layer is produced as soon as the current exceeds a certain limit (Torvén and Lindberg, 1980).

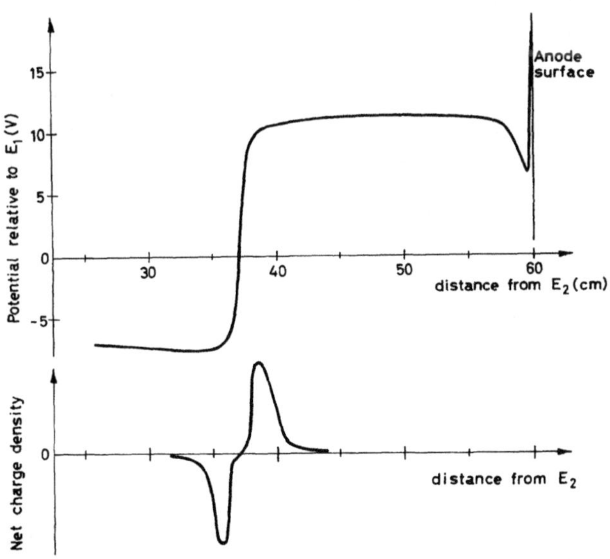

Fig. II.19. Axial potential profile at the symmetry axis (top) showing a double layer. The net charge distribution (below) gives the reason for calling it a double layer (Torvén and Lindberg, 1980).

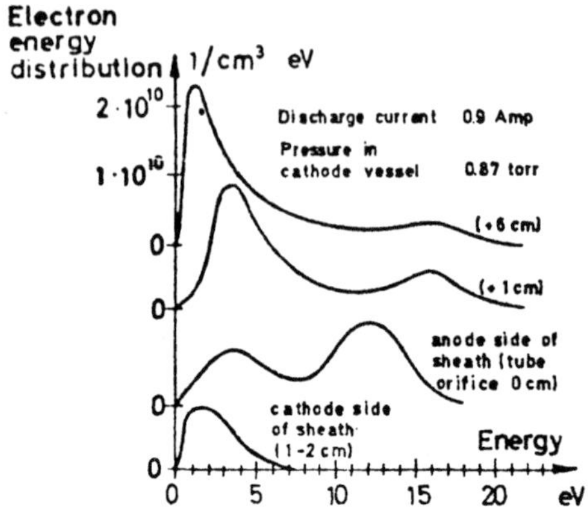

Fig. II.20. Electron energy distribution near a double layer. On the cathode side of the sheath the distribution is almost Maxwellian, with a maximum at 2 eV (bottom curve). On the anode side of the sheath with $V_D = 10$ V, this maximum is shifted to 12 eV. At the same time a new maximum begins to develop at lower energies (second curve from bottom). Further away from the sheath the low energy maximum dominates while the high energy maximum is flattened (two top curves).

of their double layer. An energy distribution of the electrons is depicted in Figure II.20 (Andersson *et al.*, 1969). On the cathode side of the double layer, the Maxwellian distribution peaks at 2 eV, but on the anode side the acceleration of the particles through the double layer causes two peaks, one at 3 eV and one at 12 eV. At larger distances from the double layer, the 3 eV signature becomes the dominant feature. Recent work in a much larger vessel confirms these results and gives a more detailed picture (Baker *et al.*, 1980).

These experiments demonstrate that the formation of double layers in the laboratory is a purely *electrostatic* phenomenon and not primarily produced by a magnetic field, which is of importance only for the confinement of the plasma. We cannot possibly describe a double layer by a magnetic field related formalism. Double layers are observed at currents so low that the magnetic field they produce is negligible (even the Larmor radius of the electrons is large compared to the extension of the plasma).

The voltage drop V_D in a double layer is of the order of ten times the voltage equivalent of the temperature, kT_e/e. In Table II.1, which shows some experimental results, eV_D/kT_e varies between 4.5 and 20, and the thickness of the double layer is 10–25 times the Debye length.

Often but not always, double layers are associated with high-frequency oscillations. The origin and significance of these has as yet not been much explored. One of their effects is that they introduce a certain spread in the energy of the particles accelerated in the double layer.

II.6.2. DOUBLE LAYERS IN THE MAGNETOSPHERE

Observations show that the auroral zone is often bombarded by nearly monochromatic electrons of, for example, 3 keV energy. It has been shown that these are accelerated in a rather narrow region at a height of about one earth radius, $R_\oplus$, above the ionosphere. The required voltage drop may be caused by the magnetic mirror effect (CE 5.1.3) or by several double layers. Both these phenomena are caused by *upward* field-aligned

Table II:1

Results from some laboratory experiments are summarized in Table II:1. This gives measured values of the ionization degree N_i, the layer voltage V_D, eV_D/kT_e (T_e is the electron temperature on the low potential side of the layer), the Debye length λ_D, and the layer thickness L. As a measure of the space charge strength within the layer = $(\lambda_D/L)\ (eV_D/kT_e)^{1/2}$ has been used. The typical space charge density in the layer is $\pm 4\gamma^2 en$ if the densities n on each side of the layer are equal (Shawhan *et al.*, 1978; Torvén, 1979).

Experiment	N_i (%)	V_D (V)	$\frac{eV_D}{kT_e}$	λ_D (10^{-3} m)	L (10^{-3} m)	$\frac{\lambda_D}{L}\left(\frac{eV_D}{kT_e}\right)^{1/2}$
Coakley *et al.*	0.01	14	14	1.5	150	0.04
Lutsenko *et al.*	100 (initially)	10^4			40	
Quon and Wong	0.01	16	4.5	1.5	35	0.09
Torvén and Babić	1	200	20	0.05	<5	>0.03
Torvén and Andersson	0.01	25	12	0.35	12	0.1
Joyce and Hubbard (num. sim.)	100		50	$\frac{L}{\lambda_D} = 40$		0.17
Torvén and Lindberg	0.1	18	9	0.2	20	0.02

currents, which tend to transport electrons from the magnetosphere *downwards* to the ionosphere. The magnetic mirror effect is due to the reflection of the magnetospheric electrons by the strongly convergent magnetic field. If the supply of electrons is not sufficient to carry the current required by the circuit, a sometimes large electric field is necessary. According to Lennartsson (1979), both the double layer and the magnetic mirror effects are active simultaneously: the magnetic mirror effect necessarily leads to the formation of one or several double layers in which the total voltage drop is produced.

II.6.3. ENERGY RELEASE IN DOUBLE LAYERS

If a double layer has been formed by a current I, energy at a rate

$$P = IV_D \tag{14}$$

is released in the double layer. This energy is mainly used for accelerating charged particles. A small fraction is usually dissipated as noise. Of course, the accelerated particles interact with the plasma and produce a number of secondary effects so that the released energy finally is dissipated as heating and radiation. Again, it should be mentioned that *there is no possibility of accounting for the energy of the particles as a result of 'magnetic merging' or of 'magnetic field-line reconnection'*, or any other mechanism which implies changing magnetic fields in the region of acceleration (II.3.3, II.5.3). In the region of the double layer, the magnetic field during the explosive transient phase is almost constant and cannot supply the required energy (of course, the secondary effects of the explosion also cause changes in the magnetic field).

II.6.4. EXPLODING DOUBLE LAYERS

As we have seen, plasma phenomena depend in a decisive way on *the properties of the whole electric circuit*. It is well-known that all electric circuits containing an inductance

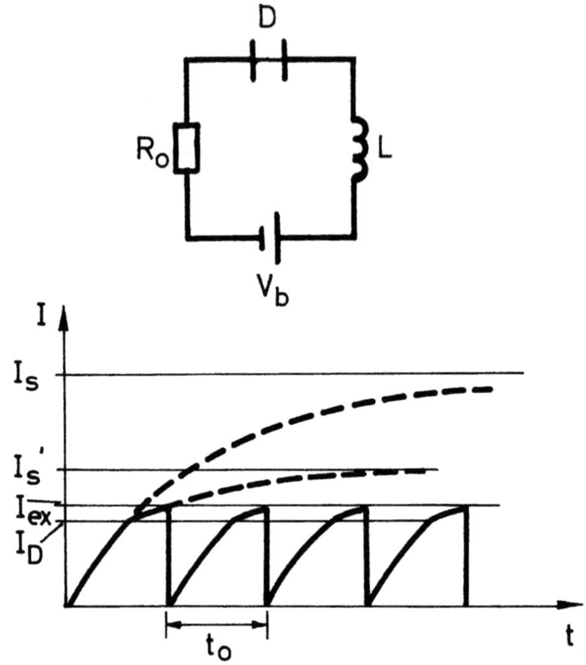

Fig. II.21. Circuit with an exploding double layer. Under certain conditions regular oscillations may be produced.

are intrinsically *explosive*. If the circuit is disrupted, the magnetic energy

$$W_L = \tfrac{1}{2}LI^2 \tag{15}$$

is released at the point of disruption. Further, it is well-known from decades of plasma investigation in the laboratory that certain types of electric double layers may become unstable in the sense that they suddenly disrupt the current. The result is that the energy W_L may be released in the double layer, and this causes an explosion. However, there may also be other plasma instabilities which cause a disruption of current, and hence an explosion (see Spicer, 1977).

There are good reasons to suppose that many of the explosive events observed in cosmic physics are produced by exploding double layers (see Carlqvist, 1969, 1972, 1973, 1979b; Boström, 1974). Examples are magnetic substorms, solar flares, and similar phenomena in 'flare stars'. The 'folding umbrella' phenomenon observed in comets has led Ip and Mendis (1976), and Mendis (1978) to suggest that a similar process may be responsible for sporadic enhancements of ionization in cometary atmospheres ('cometary aurora').

II.6.5. A CIRCUIT WITH A DOUBLE LAYER

Consider a circuit containing an e.m.f. V_b, an inductance L, a resistor R_0 and a double layer D (see Figure II.21). Suppose that the double layer is formed as soon as the current I exceeds a value I_D and that it has a constant voltage drop V_D up to the current I_{ex} when it explodes and disrupts the current.

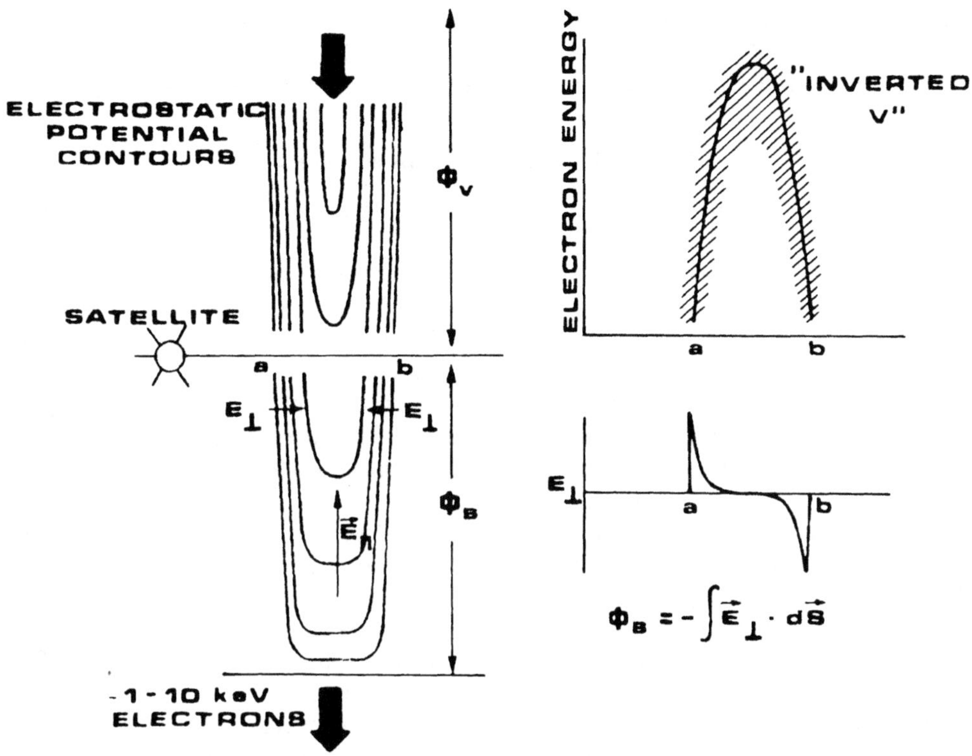

Fig. II.22. 'Inverted V' events. (Right) Observed inverted V events. (Left) Electrostatic potential distribution according to Gurnett (1972). The magnetic field is vertical. The current-carrying flux tube is 'insulated' from the surrounding plasma by a thin cylindrical shell of rotating plasma, which produces a voltage drop which equals the electrostatic drop in the layer. Compare the equipotential lines with those in Fig. II.18. Both are examples of '*cable*' formation in a plasma.

When switched on, the current increases at the rate

$$\frac{dI}{dt} = \frac{V_b - R_0 I}{L}. \tag{16}$$

Without double layers, it would reach the saturation value

$$I_s = V_b/R_0. \tag{17}$$

If $I_D < I_s$, its rate of increase for $I > I_D$ is given by

$$\frac{dI}{dt} = \frac{V_b - V_D - R_0 I}{L}. \tag{18}$$

The current will tend towards a saturation value

$$I_s' = (V_b - V_D)/R_0. \tag{19}$$

If $I_{ex} < I_s'$, the double layer will explode before the saturation value is reached.

If the circuit can be re-established, the same process can be repeated again and again with a time constant t_0 (see Figure II.21). Repetitive events are often observed (magnetic substorms, solar flashes, sometimes solar flares). In all of these cases, the repetitive properties may be due to the properties of this simple circuit. The energy which is dissipated in the explosions is derived from the kinetic energy of the plasma motions.

II.7. Field-Aligned Currents as 'Cables'

The discovery of 'inverted *V* events' by L. A. Frank has shown that electric currents in space often flow in 'cables'. See Figure II. 22.

From the observed inverted *V* events, we know that cable formation is frequent in the lower magnetosphere (up to at least one $R_\oplus$). They are not so often observed higher up, but this decrease may be observational. Direct measurements of strong electric fields have been made by Mozer *et al.* (1977). Besides electric fields parallel to the magnetic field, they find very strong fields perpendicular to the magnetic field. The shape of the equipotential surfaces is depicted in Figure II.22. Other measurements of strong electric fields have been made both with satellite (Maynard, 1978; Smiddy *et al.*, 1977) and rocket (Marklund *et al.*, 1979).

Accelerating regions of the type that we have described have been observed with barium cloud experiments at altitudes of about one Earth-radius (Haerendel *et al.*, 1978; Wescott *et al.*, 1976; Shawhan *et al.*, 1978).

If the current-carrying field tube has the same voltage as the environment below the double layer, there must be a lateral voltage gradient above the layer. This produces a rotational motion of the plasma (but should not be depicted as a motion of magnetic field lines!) around the current-carrying flux tube. In this way, *the filamentary current is electrically insulated from the surroundings in a way similar to a current in an electric cable* located in the ocean and carrying current through a low resistance metal wire. The wire is insulated from the conducting water surrounding it by a plastic cylinder in which the electric field is similar to the radial electric field surrounding the field aligned current in the magnetosphere.

In the same way as two high power transmission cables connect a generator and a 'consumer', a pair of plasma cables may connect a generator and a 'consumer'. The generator often consists of a plasma moving with a velocity component perpendicular to the magnetic field **B** and hence generating an e.m.f.

$$V_b = \int \mathbf{v} \times \mathbf{B}\, \mathrm{d}\mathbf{s} \qquad (20)$$

where the integral is taken between the ends of the two cables. The consumer may be a double layer accelerating charged particles which later produce light or synchrotron radiation. The consumer may also be a 'motor' which sets plasma in motion in a distant region. As an example, in the auroral circuit, the generator is located in the solar wind and the 'consumers' are double layers accelerating high energy particles which later illuminate the night sky in the auroral zone and/or a 'motor' which produce the sun-ward drift in the magnetosphere (see Figure II.17 and Alfvén, 1979).

The observed existence of cable-like plasma configurations motivates us to draw electric circuit diagrams for electromagnetic phenomena in space, and to discuss them with the help of electrotechnical terminology. This method will be extensively used, especially in Chapter III. It is obvious that it should be regarded as a first approximation to a more complicated situation. Great care is necessary in order to determine to what extent it may be misleading.

II.8. An Expanding Circuit

There is also another important way in which the energy $\frac{1}{2}LI^2$ of the circuit can be dissipated. If an electric current flows in a circuit, it exerts an electrodynamic pressure such that the circuit loop has a tendency to expand. In cosmic plasmas, this pressure is usually balanced by other forces (e.g., other electromagnetic effects, gravitation, gas pressure). If the current increases above a certain value its electromagnetic pressure may be so large that the balancing effects do not suffice and the current loop may explode. An example of this is the rising prominences. In this case part of the circuit energy is converted into kinetic energy.

II.9. Different Types of Plasma Regions

What has been said above might be summarized and elaborated in the following way: it seems that at least in low density cosmic plasmas there are the following three different types of regions.

II.9.1. 'PASSIVE' PLASMA REGIONS

These regions may transmit different kinds of plasma waves and high energy particles. There may be transient currents, perpendicular to the magnetic field, changing the state of motion of the plasma, but not necessarily associated with strong electric fields and currents parallel to the magnetic field. A plasma of this kind fills most of space.

If in an initially homogeneous plasma, the parameters in one region are changed above a certain limit, the plasma often reacts by setting up a discontinuous interphase. Quite generally, instead of a continuous space variation of the parameters, the plasma 'prefers' to produce a discontinuity interphase separating two almost homogeneous regions. This may sometimes be the origin of *double layers*, taking up voltage differences, and *current sheets*, separating regions of different magnetization, temperature, and density.

II.9.2. 'ACTIVE' PLASMA REGIONS

Hence, besides the passive plasma regions, there are also active plasma regions where filamentary and sheet currents flow (Alfvén, 1977). Since they transfer energy and produce sharp borders between different regions of passive plasmas, they are of decisive importance to the overall behavior of plasmas in space. This is true, even if their total volume is small, as is often the case. We refer to these two different types of active plasma regions as *plasma cables* and *boundary current sheets*.

II.9.3. PLASMA CABLES

Plasma cables seem to be reasonably stable formations which can be considered as structures important for the understanding of plasma phenomena. (Of course, their interior structure should be described by classical theory.) The plasma cables are either filaments or 'flattened filaments' (sheets with limited extent). They carry an electric current parallel to the magnetic field, and this is what gives them their properties. The cables are often very efficient in transferring electromagnetic power from one region to another. They are embedded in passive plasmas, which have essentially the same properties in all directions around the cables. They are 'insulated' from their surroundings by a thin cylindrical electrostatic sheath (or double layer) which reduces the interaction with its exterior. In the magnetosphere and upper ionosphere, the density in the cable is sometimes lower than the surrounding passive plasma (Block and Fälthammar, 1968). In other cases, the density in the cable may be much larger than the surroundings because ionized matter is pumped into the cable from the outside. By selectively doing so, the chemical composition in the cable may differ from that of its exterior (Marklund, 1978, 1979). Besides the cylindrical electrostatic sheath, there are often longitudinal *double layers*, in which a considerable part of the power which the cable transmits may be converted into high energy particles. The double layers sometimes explode, and this produces excessively high energy particles. Then the current in the cable may be disrupted and switched into another circuit (Boström, 1974, 1975). Besides in the laboratory, such plasma cables have been discovered (or are likely to exist) in several places:

(a) *The auroral current system* consists of plasma cables, sometimes filaments and more frequently sheets (possibly associated with auroral rays). The inverted *V* events are likely to be produced by cables (Block, 1978). In the auroral zone such cables transfer energy from the equatorial plane to the auroral zone, and double layers are likely to be the direct cause of some types of luminous aurorae (see III.2).

(b) *Solar prominences*. The electromotive force is due to motions in the photosphere in combination with the photospheric magnetic fields. The prominences are often filamentary, but sometimes they are sheet-like (quiescent prominences). Double layers are often produced, and when they explode, they cause solar flares. *Spicules* are probably small scale versions of the same phenomenon (see III.8).

(c) *Coronal streamers and polar plumes*. As we expect that the large heliospheric current system should connect with the Sun in the polar regions and in the equatorial region, it seems likely that these phenomena are basically similar to the solar prominences. The main difference is that the prominences are produced in a photospheric circuit, whereas the streamers and plumes are caused by the large heliospheric current system (III.4.3).

(d) The filamentary structure in *cometary tails* (Mendis, 1978; and see Figure II.6), and in *interstellar* nebulae (Figures II.7–10) may be explained in a similar fashion. *Stellar flares* are probably due to the same mechanism as solar flares. It has also been suggested that the transfer of energy to the *double radio sources* from the galaxy, which is normally located halfway between them, is due to a plasma cable of this type (III.4.4).

II.9.4. IONOSPHERIC PROJECTION OF ACTIVE AND PASSIVE PLASMA REGIONS

The drastic difference between active and passive plasma regions is visualized by a look at the ionosphere. The auroral zones are those regions of the ionosphere which magnetically connect to the active region in the magnetosphere. The rest of the ionosphere connects with passive plasma regions, and hence receives very little energy from higher altitudes.

It is tempting to identify the different auroral patterns with the different active structures in the magnetosphere: auroral rays may derive from active filaments, arcs from sheets, draperies from perturbed sheets. However, there are quite a few rather complicated phenomena in the transitional region between the magnetosphere and the ionosphere. Further, there are horizontal currents in the upper ionosphere, which may also produce luminous effects. Hence, the relation between the structure of the active plasma in the magnetosphere and the visually observed aurora is very complicated. Finally, it should be remarked that filaments are more easily seen at large distances, whereas sheets are more easily detected from spacecrafts.

II.9.5. BOUNDARY CURRENT SHEETS

In striking contrast to the current layer we have discussed above, there is another type of current layer which forms boundaries between two regions of plasma possessing drastically different parameters. The basic difference is that the cables we have discussed above carry currents so small that their magnetic field is negligible compared to the exterior field, whereas in the boundary current sheets, the current is large enough to change the surrounding field considerably. In the magnetosphere, there are three such layers (Heikkila, 1975; Heikkila and Block, 1977):

(a) *The magnetopause*, which is a very thin current layer (a few cyclotron radii thick), separates two regions which often have opposite magnetic fields (directed northward and southward). The other parameters (pressure, density) are not necessarily very different but may be so.

In many cases, a spacecraft passing the region of the magnetopause observes more than one magnetic discontinuity. This is usually interpreted as an indication that the magnetopause moves very rapidly in and out. The data can possibly also be interpreted as a 'folding' of the magnetopause current layer. Auroral draperies are often observed to fold so that they become *S*-shaped. This process may continue, resulting in a triple layer. Similarly, the heliospheric boundary layer is known to be wavy. By analogy, it is quite reasonable that the magnetospheric current layer is folded so that a spacecraft traverses it three times. A model of this has been given by Smith (1979). However, the layer movement interpretation seems more probable.

(b) *The neutral sheet in the tail* also separates two regions with essentially opposite magnetic fields (directed towards and away from the Earth). Other parameters are usually similar. It is also believed that the magnetic field frequently has a small northward magnetic field superimposed.

(c) Further, there is a *current layer in the front* of the magnetopause which separates a region of very weak magnetic field in the solar wind from a much strong field (three times or more) between the front current and the magnetopause. This current layer is usually referred to as the '*bow shock*'. It transforms a large part of the solar wind kinetic energy into electromagnetic energy (see III.6).

In the *heliosphere*, there is a very extended current layer in the *equatorial* plane, which displays a north–south oscillation and separates an inward magnetic field from an outward directed field (Figure III.7) (Alfvén, 1977; Akasofu *et al.*, 1979). Current layers similar to those in the magnetotail and heliosphere have been discovered in the Jovian magnetosphere as well as other magnetospheres, and are theoretically inferred in the tails of comets. Recently such a layer has also been observed at the antisolar side of Venus (Gringauz, 1980).

The Earth's magnetotail current sheet is likely to produce explosive double layers (Kan *et al.*, 1978) which cause magnetic substorms. Whether other current sheets have the same property is not known. There are a number of theories of each of these current layers, but unfortunately, a general theory covering all these very important phenomena still seems to be lacking.

The discovery of these layers (the first one, the magnetopause, discovered by Cahill) appears to many as the most sensational of all space research discoveries (although Dungey (1961, 1964) had predicted them to some extent). It is especially surprising that these current layers exhibit such a remarkable permanence, if not stability. The existence of such layers changes our views of the structure of space plasmas (Fälthammar *et al.*, 1978). As such layers are of course likely to exist also in regions where spacecraft have not yet made *in situ* measurements, we must drastically revise our views of the structure of space. The existence of such boundary layers leads to the conclusion that space everywhere (interstellar space and intergalactic space) has a *cellular structure*. We shall discuss the consequences of this in the next section.

II.10. Cellular Structure of Space

As we have seen in the preceding chapters, a number of interfaces have been discovered which separate regions of different magnetization, density, temperature, electron velocity distribution, and even chemical composition. Examples are: the magnetopause and magnetotail sheets, the heliospheric equatorial sheet, similar sheets in the Jovian magnetosphere, and possibly also in the cometary tails.

These sheets are caused by electric surface currents. They are sometimes very thin (down to about ten times the ion Larmor radius). It is almost impossible to detect them from a distance. A spacecraft usually sees no indication of such a sheet until it actually passes through it.

As it is unlikely that cosmic plasmas have such properties only in those regions which are accessible to spacecraft diagnostics, it is legitimate to conclude that space in general has a '*cellular structure*', although this is almost impossible to observe unless a spacecraft penetrates the 'cell walls' (current sheets). This means that in distant regions, we cannot hope to detect the cell walls directly. Nor can we tell the size of the cells. It is unpleasant to base far-reaching conclusions on the existence of a structure which we cannot detect directly. But the alternative is to draw far-reaching conclusions from the *assumption* that in distant regions, the plasmas have properties which are drastically different from what they are in our own neighborhood. This is obviously far more unpleasant than our inability to detect distant 'cell walls'. Hence, a thorough revision of our concept of the properties of interstellar (and intergalactic) space is an inevitable consequence of recent magnetospheric discoveries. As we shall see in IV.9 and VI, this must also change our approach to cosmology.

II.11. Fine Structure of Active Plasma Regions

The thickness of active plasma regions is a matter of controversy. The auroral zone has an extension of a few degrees (= some 10^5 m). However, the thickness of an auroral arc or ray is often of the order 10^3 m, sometimes perhaps even smaller. Measurements of the thickness of magnetopause current layers are often interpreted as indicating a thickness of a few orders of magnitude more than the ion cyclotron radius. However, there are also many measurements giving thicknesses of only a few cyclotron radii. A possible interpretation is that active plasma regions have a *fine* structure of the order of a few cyclotron radii which is the basic element determining the fundamental properties of active plasmas.

Although solar prominences are supposed to be 10^6 or 10^7 m in diameter, there are many cases where a filamentary fine structure has been observed. This indicates that a prominence should be considered as a 'rope' consisting of a multitude of small wires. It is also possible that the smallest observed filamentary structure in the sun (some 10^5 m) consists of still smaller filaments, perhaps as small as the size of structures observed in the ionosphere (10^3 m).

It seems doubtful whether a deeper understanding of the properties of a plasma is possible before we are able to study its fine structure. Hence, the study of active plasma regions according to the principles we have discussed may very well be an important step forward, but should not necessarily be regarded as the final step.

An important discovery of small scale current ropes (diameter 10^4 m) was recently made by the Pioneer Venus Orbiter (Russell *et al.*, 1979; Elphic *et al.*, 1979). The Cytherean ionosphere is permeated by a large number of twisted magnetic filaments of the same kind as recently studied theoretically by Marklund (1978). They differ from the auroral rays of our ionosphere in the respect that they are *magnetically* contracted. It is possible that the observed phenomenon is the long sought-for filamentary structure which should be the basic constituent of filamentary currents in prominences, coronal streamers, perhaps cometary tails and cosmic plasmas in general.

We have described here three typical regions of plasma: *passive plasmas*, which seem to be the most common type in the magnetospheres of the planets and the Sun; *cables*, which are found in the auroral zones; and *boundary current sheets*, of which the magnetopause is the best studied example. However, it is quite likely that there are many cases which are intermediate between these three types. A detailed analysis is needed before it is possible to decide this.

CHAPTER III

CIRCUITS

III.1. Importance of Electric Current Models

As we have seen in Chapter II, it is advantageous to translate the traditional field description of plasma phenomena in space into a current description. This gives a deeper understanding of those phenomena in the following respects:

(1) The circuit representation demonstrates the importance of boundary conditions which have often been forgotten.

(2) By studying the electromotive forces driving the current and the regions of dissipation, the energy transfer from one region to another is more easily understood.

(3) Certain types of important current-produced phenomena, including the formation of double layers, are difficult to understand without accounting for the current explicitly.

As an example, by simply translating the field description of the heliosphere into a current description, we can demonstrate that there must be currents along the solar axis (which had not been derived from the field description), and that coronal streamers and polar plumes are likely to be due to filamentary currents (III.4). We shall also make a similar translation of the magnetic field description of the magnetosphere into an electric current description (III.6) which makes it possible to understand the transfer of energy from the solar wind to the magnetosphere. By a similar method, we can account for the transfer of energy from photospheric activity to the solar wind (III.9). This will make it possible to understand the whole sequence of processes which transfer energy from the Sun to the magnetosphere and ionosphere (III.10).

III.1.1. PARTICLE DESCRIPTION

In contrast to the usual hydromagnetic description, we shall use the *particle description* exclusively (see Chapter II). Neglecting collisions, gravitation, and radiation pressure the changes in the state of motion of a particle are entirely caused by the electromagnetic forces $F_E = e(\mathbf{E} + \mathbf{v} \times \mathbf{B})$ which act on the particle. Hence, if a particle stream meets an obstacle, it is deviated only by the force F_E. *The effect of the obstacle is due to the electric and magnetic fields it produces.* What this means will be explained by some examples discussed in III.1.3.

Further, because the Spitzer formula for conductivity is derived under the assumption of a small mean free path, it is not applicable. If a voltage difference is applied parallel to the magnetic field, or along a neutral line, we expect a current to flow. The voltage–current relation does not, however, obey Ohm's law (Block and Fälthammar, 1976). Double layers and mirror-related voltage drops are more important. Furthermore, a plasma has many different types of instabilities, and is normally in a 'noisy' state. This does *not* mean that it is *turbulent* (IV.4). The sloppy use of the term 'turbulent' has caused, and is causing, much confusion. A spacecraft moving through a plasma often registers rapid fluctuations, but sometimes these are due to the filamentary structure of plasma, and sometimes due to waves of different kinds. However, there is no certain indication that *anywhere* in space there is very much (large scale) turbulence in the proper sense of the word. This is important, for the reason that real turbulence produces mixing, and there is no certain evidence for a high degree of mixing in space plasmas. On the contrary, a separation of elements often takes place (IV.3).

III.1.2. DIFFERENT TYPES OF ELECTRIC CURRENTS

Three main types of electric currents are of importance:
(a) Field aligned currents $i_{\parallel}$ parallel to B.
(b) Currents i_n along neutral lines or surfaces where $B = 0$.
(c) Drift currents $i_{\perp}$ perpendicular to B.

(a) *Field-aligned Currents*

One of the features of parallel currents is that in a low density plasma, they do not obey Ohm's law. They may produce double layers, causing large voltage drops and noise (II.6). Their properties have been demonstrated primarily by experiments, but magnetospheric research has now shown that they are of crucial importance for the understanding of space plasmas. They are complicated, and present theories are essentially semi-empirical.

Another important feature of the field-aligned currents is that they are often filamentary. There is good reason to believe that observed filamentary structures in space are produced by field-aligned currents (II.4).

(b) *Currents Along Neutral Lines or Surfaces*

There are a number of theoretical treatments of neutral line (or surface) currents, but as these currents are less well studied in the laboratory we do not know to what extent the theories concerning them are valid. In some respects, neutral line currents may be similar to field-aligned currents. Like field-aligned currents, they may be likely to form exploding double layers. Experiments by Ohyabu *et al.* (1974) seem to confirm this (although their interpretation is different).

(c) *Drift Currents*

The perpendicular drift of a charged particle obeys the guiding center equation (see CE, p. 37):

$$\bar{u}_{\perp} = (\mathbf{F}_E + \mathbf{F}_B + \mathbf{F}_g + \mathbf{F}_i + \mathbf{F}_0) \times \mathbf{B}/qB^2. \tag{1}$$

The drift motion consists of the following components, deriving from the different terms in Equation (1)

the electric field drift:	$\mathbf{F}_E = q\mathbf{E}$
the magnetic gradient drift:	$\mathbf{F}_B = -\mu \operatorname{grad} B$
the gravitation drift:	$\mathbf{F}_g = m\mathbf{g}$
the inertia drift:	$\mathbf{F}_i = -m \dfrac{\mathrm{d}\mathbf{u}}{\mathrm{d}t}$

where m = mass of the particle, $\mu = P_{\perp}^2/2\gamma mB$ = magnetic moment (in the non-relativistic limit $\mu = mv_\perp^2/2B$), and g = gravitational force. Finally, $\mathbf{F}_0$ represents other forces deriving from, e.g., radiation pressure or collisions. The 'perpendicular' kinetic energy $\frac{1}{2}mv_\perp^2$ may represent the 'perpendicular temperature'

All of these drifts except the electric field drift depend on the sign of the charge and hence *produce currents*. The gravitation drift depends on m and hence can produce a chemical separation of atoms with different masses. This is also true for the inertia drift. The magnetic gradient drift may also separate different species of ions if their perpendicular temperature differs.

The inertia drift u_i is important because *it transfers kinetic energy into electromagnetic energy, and vice versa*. If for a first approximation we put the plasma temperature equal to zero so that $\mathbf{F}_B$ is zero, the inertia drift current density i_i is given by

$$\mathbf{i}_i = \sum q n \mathbf{u} = \sum \frac{nm\mathbf{B}}{B^2} \times \frac{\mathrm{d}\mathbf{u}}{\mathrm{d}t} \tag{2}$$

(n = particle density, m = particle mass)

where the sum is taken over the different species of ionized particles. This '*inertia current*' flows perpendicular to B, which means that it exerts the force per unit volume

$$\mathbf{f}_i = \mathbf{i}_i \times \mathbf{B} = \rho \frac{\mathrm{d}\mathbf{u}}{\mathrm{d}t} \tag{3}$$

(where $\rho = nm$ is the plasma density) which is necessary to accelerate the plasma at the rate $\mathrm{d}\mathbf{u}/\mathrm{d}t$.

III.1.3. TRANSFER OF ENERGY BETWEEN A CIRCUIT AND A MOVING PLASMA

In order to study the energy transfer between *kinetic energy* $W_k = \frac{1}{2}Mu^2$ of a plasma and electromagnetic energy of a circuit, which will be referred to as *circuit energy* W_c, we consider the following three simple cases.

Model (a)

In a homogeneous magnetic field B, strong enough to make the guiding center approach valid, a thin slab of plasma with thickness Δx is accelerated in a condenser A called the *plasma gun* (see Figure III.1(a)). The distance between the condenser plates is Δy and the breadth in the magnetic field direction is Δz. The extension of the condenser in the direction of the x-axis is large enough to ensure that the plasma reaches its limiting velocity in the gun. The plasma then moves through a *drift tube*, where there is no exchange of energy, to another condenser, called the *deceleration region D*.

The gun is energized from a battery with the voltage V_0, which is connected to the plasma gun through a resistance R_A via a double switch S which is closed when the gun is activated. The deceleration condenser is connected to a resistor R_D. The whole device is symmetric with respect to ground. The plasma slab contains a fully ionized homogeneous plasma with density ρ and mass $M = \rho\Delta x\Delta y\Delta z$. Further, it is assumed that electrode phenomena in A and in D can be neglected (this means that all electrodes can emit ions and electrons without limitation, but the increase in mass M because of this can be neglected).

When the plasma gun is switched on, a current I accelerates the plasma with a force

$F = \Delta y I B$ to a velocity u. We denote the voltage across the plasma gun by V and have

$$V = V_0 - IR_A. \tag{4}$$

Further,

$$u = \frac{V}{B\Delta y} \tag{5}$$

and the power P delivered to the plasma is

$$P = IV \tag{6}$$

Combining Equations (2) and (5), we obtain for the total current through the surface area $\Delta x \Delta z$

$$I = \frac{nm\Delta x \Delta z}{B} \frac{\mathrm{d}}{\mathrm{d}t} \frac{V}{B\Delta y}. \tag{7a}$$

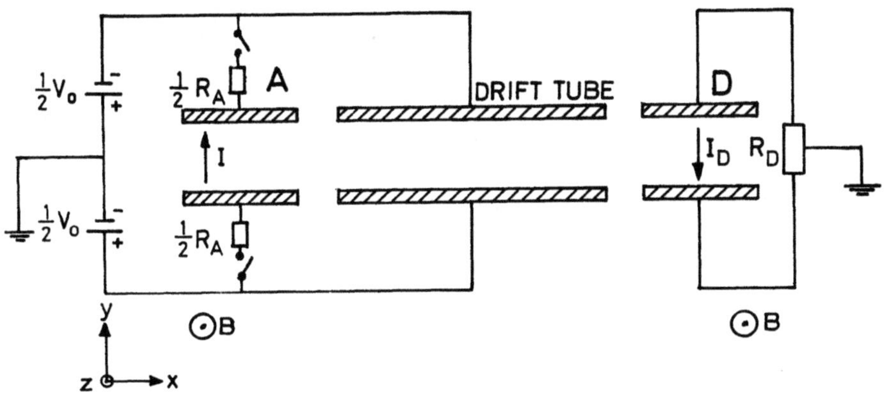

Fig. III.1(a). Circuit of model (a). Magnetic field B is homogeneous. The plasma is accelerated in region A by the force $F = \Delta y I B$ to a velocity u. After passing the drift tube it enters the region D. Here the plasma decelerates causing an e.m.f.

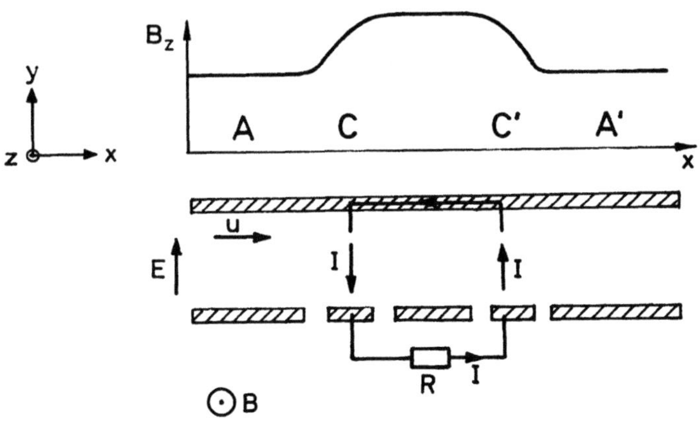

Fig. III.1(b). Circuit of model (b). The magnetic field B is larger in CC' than in A and A'. When the plasma enters and leaves the stronger field the gradients cause currents as shown. The models (a) and (b) illustrate the basic mechanisms for exchange between plasma kinetic energy and circuit energy.

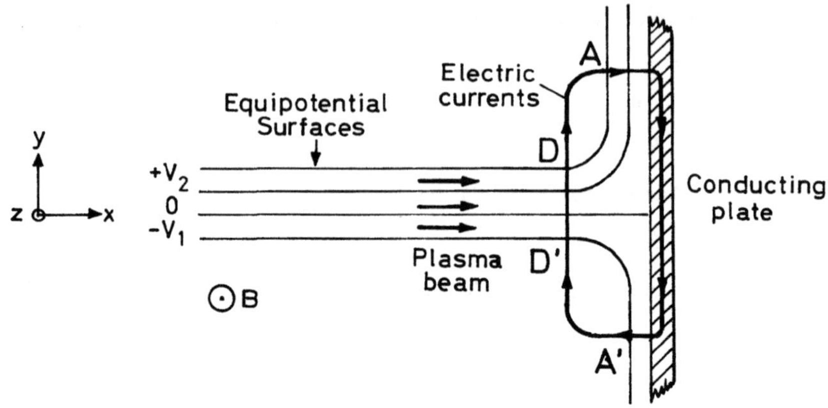

Fig. III.1(c). Circuit of model (c). Homogeneous magnetic field. A plasma beam shot towards a conducting plate is deviated because the plasma drifts approximately along the equipotential surfaces. The deceleration at DD' causes an inertial current which closes as depicted, and causes acceleration in the regions A and A'. Hence the deviation of a low density magnetized plasma beam shot towards a conducting obstacle is neither caused by a collision with the obstacle nor by a pressure gradient. Only electric and magnetic fields act on the particles.

Incorporating Equation (4), this may be written as

$$I = -t_A \frac{\mathrm{d}I}{\mathrm{d}t} \tag{7b}$$

with

$$t_A = R_A \frac{S}{\Delta y} \frac{\rho}{B^2} \tag{8}$$

where $S = \Delta z \Delta x$ is the effective surface of the electrodes.

Integrating, we have

$$I = I_0 \, \mathrm{e}^{-t/t_A} \tag{9}$$

with

$$I_0 = V_0 / R_A \, . \tag{10}$$

As the capacitance C of a plane condenser is

$$C = \epsilon_0 \frac{S}{\Delta y} \tag{11}$$

(where ϵ_0 is the dielectric constant in vacuum)

We can write Equation (8) as

$$t_A = R_A \frac{C}{\epsilon_0} \frac{\rho}{B^2} \tag{12}$$

$$= R_A C \epsilon / \epsilon_0 \tag{13}$$

where the 'apparent dielectric constant' ϵ is

$$\epsilon = \frac{1}{\mu_0 V_A^2} \tag{14}$$

where
$$V_A = (\mu_0 \rho)^{-1/2} B \tag{15}$$
is the hydromagnetic velocity.

When the plasma slab enters the decelerating device, the current will suddenly rise to $I_0 = V_0/R_D$ and then decrease according to Equation (7b), but with t_A replaced with $t_D = R_D C\epsilon/\epsilon_0$.

If the decelerating device is very long, the plasma will deliver all its kinetic energy to the circuit. If it is so short that the plasma leaves before V has gone to zero, the plasma releases only part of its kinetic energy to the circuit, and proceeds further with reduced speed.

Comparing the state of the plasma in the drift tube with the state before acceleration, it is found that guiding centers of the ions have been displaced a distance y_p in the direction of the electric field $E = V_0/\Delta y$ which is large enough to give them their kinetic energy. This means
$$y_p E = \tfrac{1}{2} m \frac{E^2}{B^2} \tag{16}$$
or
$$y_p = m \frac{E}{2B^2}. \tag{17}$$

When they are decelerated again, the displacement goes down to zero. This displacement is equivalent to the displacement in the dielectric of a condenser. The change in y_p gives the inertia current. The electrons are displaced in the opposite direction, but because of their smaller mass, this displacement is unimportant.

Model (b)

Consider a plasma drifting in a magnetic field $B_z(x)$ under the influence of an electric field E_y, deriving from two condenser plates with unit length $\Delta x = 1$ and unit separation $\Delta y = 1$. The drift velocity u is given by Equation (1). If n is the density of the electrons and of the ions, which have masses and magnetic moments m_e, m_i, and μ_e, μ_i, the current density is
$$\mathbf{i} = ne(\mathbf{u}_i - \mathbf{u}_e) = \frac{n}{B^2} \mathbf{B} \times \left[(\mu_i + \mu_e) \operatorname{grad} B + (m_i + m_e) \frac{\mathrm{d}\mathbf{u}}{\mathrm{d}t} \right] \tag{18}$$
from which we can derive the energy delivered per m^2 sec $P = Vi$
$$P = \nu \frac{\partial}{\partial x} (w_k - w_t) \tag{19}$$
where ν is the flux of ion-electrons, w_k the kinetic, and w_t the thermal energy per ion-electron.

If plasma originating in region A (Figure III.1(b)) enters a region C with increasing magnetic field, it can react in two different ways:

(1) The velocity is constant because of the inertia. Hence, E increases proportionally to B. If B returns to its initial value, E does the same.

(2) If E is kept constant, the velocity will decrease as B^{-1}. This requires a transfer of kinetic energy into circuit energy. In hot plasma there is also a transfer into internal

thermal energy at the rate $\nu(\partial w_k/\partial x)$. Assuming that electrode losses are negligible, a circuit as in Figure III.1(b) can transfer and reinject the energy into the plasma in another region C' where B decreases. Hence, in this way the plasma beam can pass a hump in B without any lasting change in its kinetic energy. This holds also for a hot plasma under the condition that there is no velocity scattering between $v_\perp$ and $v_\parallel$. If that should be the case, the plasma will have been permanently heated at the expense of the kinetic energy.

If the circuit transfer causes losses (e.g., in a resistance R), a smaller energy is reinjected, and the plasma flow will be slowed down (for applications, see III.6). Both in models (a) and (b) the conditions become slightly more complicated if we treat a continuous flow of plasma, because contractions of the beam in the y direction will be of importance. However, the basic principles of energy transfer remain the same.

Model (c)

As a third application we consider a homogeneous magnetic field in which a homogeneous plasma beam, limited by equipotential surfaces $+V_1$ and $-V_2$, is shot towards an infinite conducting plate at zero potential. What happens depends on the voltages V_1 and V_2, the density of the beam and the electric field at large $+y$ and $-y$ values. Suppose first that the beam intensity is so low that the currents it produces do not change the magnetic field, and that the equipotential lines have the shape depicted in Figure III.1(c).

To a first approximation both the ions and the electrons move along the equipotential lines. However, as the electric field decreases at the bend and increases again at larger positive and negative y values, inertia currents are produced due to the deceleration at D and D' and acceleration at A and A'. This produces the current system depicted in the figure (compare model (b)).

If we increase the beam intensity this current system will produce an increased magnetic field at the bends of the beam. This will slow down the beam still more so that the inertia currents will be larger and more concentrated. At large enough beam densities an inertia surface current (a 'shock') may be produced.

The transfer of energy from the deceleration region to the acceleration regions by the circuit follows the pattern of model (b).

An interesting conclusion from our model is that a low density and low temperature beam shot towards a conducting obstacle is deviated *exclusively because the obstacle perturbs the electric field* associated with the beam. If the beam intensity is low no appreciable magnetic field perturbation is produced, and if the beam temperature is small no appreciable pressure gradient is produced either, and still the beam is deflected.

This problem belongs to the category which should be treated by the electric current formalism (cf. Figure I.4).

III.2. The Auroral Circuits

Early magnetic observations in the auroral zone demonstrated the existence of strong east–west horizontal currents in the ionosphere. Birkeland concluded that they were likely to close through vertical currents, which he tried to explore by observations and with the help of his terrella experiments. However, Chapman claimed that because the earth was surrounded by 'vacuum', there could be no charge carriers in its environment, and hence no vertical currents. He and Vestine constructed a current system which was

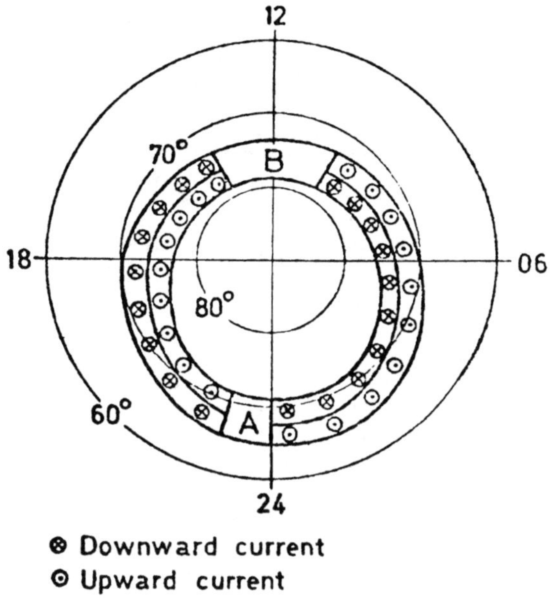

Fig. III.2. Vertical currents in the auroral zone observed by Zmuda and Armstrong (1974). On the evening side the downward currents flow at a lower latitude than the upward currents; on the morning side the directions are reversed.

confined to the ionosphere. Because of Chapman's great authority, attempts to develop the Birkeland ideas received little attention, until space measurements by Zmuda and Armstrong (1974) decided the half-century-long controversy in favor of the Birkeland approach.

Through space measurements it has been shown that there are regions of upward and downward currents separated in latitude by only about 5° (Figure III.2). On the evening side the upward current region has a higher latitude than the downward current region. On the morning side the reverse is true. Measurements of the ionospheric currents which close the Birkeland circuits have demonstrated that on the evening side this is done through northward currents and on the morning side through southward currents.

Another current system, mainly investigated by the Johns Hopkins group (cf. Potemra, 1979; Potemra *et al.*, 1980; Krimigis *et al.*, 1978; Bostrom *et al.*, 1976) also includes longitudinal currents and is depicted in Figure III.3. These west–east currents flow in sheets, but the current density is far from constant. The sheets are sometimes divided into a number of filaments, often with elliptical cross-section.

The currents are highly variable, and their structures are often extremely complicated. In order to understand the basic mechanisms, we discuss here some simple models (reserving the large scale current model for III.6). We represent the observed current system by three circuits, two with currents flowing in the meridional plane (ACI) (Figure II.17), and one representing the longitudinal current (ACII) (Figure III.4). As seen from the day side, the current in ACI flows in the clockwise direction on both the evening side and the morning side.

Having clarified the geometry of the circuits, we have to find the e.m.f. which drives

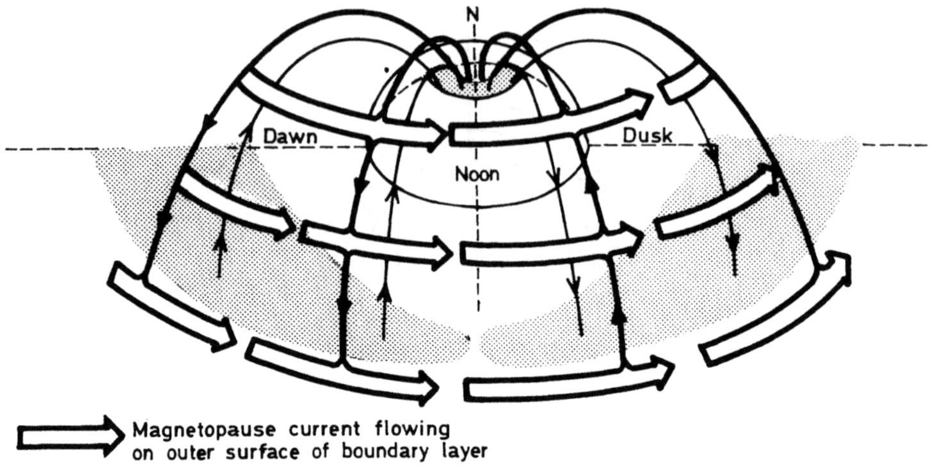

Fig. III.3. Current system according to Potemra (1980). The west–east current sheet is represented by large arrows. This is located outside the current system of Figure III.2, of which only the outer currents are indicated.

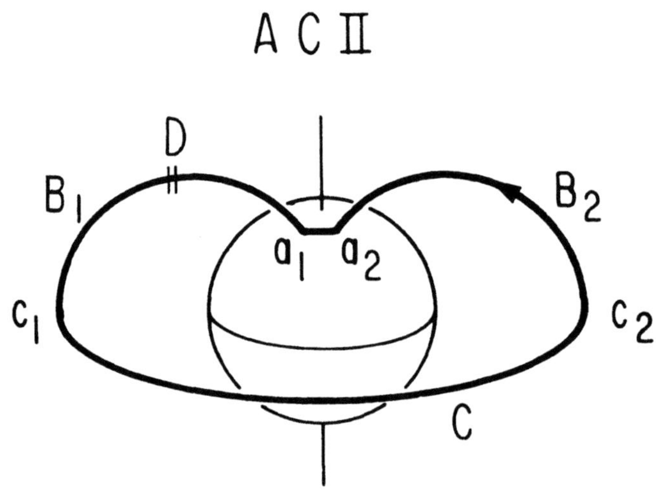

Figure III.4. Auroral circuit II. The circuit is similar to ACI (Figure II.17) with the current-carrying field lines B_1 and B_2 at the same latitude but different longitudes. Curve c_1c_2 is part of a circle in the equatorial plane; a_1a_2 is part of the auroral zone. Compare Figure III.3.

the current, further the circuit elements R and L, and the double layers, if any. (We use the term 'double layer' for the field-aligned voltage drop in the lower magnetosphere even if, perhaps, the magnetic mirror effect may be partly responsible for it (see II.6.2).)

The e.m.f. V_b is given by

$$V_b = \oint \mathbf{v} \times \mathbf{B} \cdot d\mathbf{s} \tag{20}$$

taken over the whole loop. As the integral is zero along the lines of force B_1 and B_2 (see

Figure II.17), we have $V_b = \int_{a_1}^{a_2} + \int_{c_2}^{c_1}$. We use a coordinate system following the Earth's rotation. Hence, $\int_{a_1}^{a_2}$ derives exclusively from winds in the ionosphere. Neglecting this contribution we find

$$V_b = \int_{c_2}^{c_1} (\mathbf{v} - \omega \mathbf{r}) \times \mathbf{B}\, \mathrm{d}s \tag{21}$$

where $\mathbf{v} - \omega \mathbf{r}$ is the difference between the plasma velocity and the synchronous velocity. Energy is delivered to the auroral circuit at the expense of the kinetic energy of the plasma in the equatorial plane. This transfer of energy is similar to that in the decelerating region D of model (a) in III.1.3.

As a numerical example, a value can be calculated for an auroral latitude λ of 67°. A magnetic line of force intersects the equatorial plane at a distance from the origin

$$r_e = \frac{R_\oplus}{\cos^2 \lambda}\,. \tag{22}$$

The magnetic field strength at this point is

$$B = \frac{\mu_0}{4\pi} \frac{\mu_\oplus}{r_e^3} \tag{23}$$

where $\mu_\oplus$ is the Earth's magnetic moment. Since $\mu_\oplus = 8.2 \times 10^{22}$ Am^2 and $R_\oplus = 6.4 \times 10^6$ m we reach a value of 1×10^{-7} T. With $\int \mathrm{d}s = 10^8$ m a voltage of 3 kV requires $(v - \omega r) = 0.3$ km s^{-1}.

A rough estimate of the *inductance* L can be made by remembering that for a circular loop with radius a and conductor radius b

$$L = \mu_0 a \left(\frac{1}{4} + \log \frac{8a}{b} - 2 \right) . \tag{24}$$

If $a = 10^7$ m and $b = 5 \times 10^3$ m, we obtain $L = 30$ H.

The Ohmic resistance R of the circuit, if calculated according to Spitzer's formula, is usually very small except in the ionosphere, but R should include all other losses, which are difficult to calculate.

In case there is no double layer (II.5 and II.6.5), the current will increase to a saturation value I_s. If an explosive double layer is formed before this value is reached, the current will be disrupted. Successive repetitions of the same process may be produced (see II.6.5 and Fig. II.21).

III.3. Rotating Magnetized Body Surrounded by a Plasma

As in the auroral current system of Figure II.17 currents flow perpendicular to the magnetic field both at $a_1 a_2$ in the ionosphere and at $c_1 c_2$ in the plasma in the equatorial plane, it is obvious that they transfer angular momentum between the central body and the surrounding plasma.

Hence, the exploration of the auroral current system is important for the clarification of a basic problem in astrophysics, viz., the interaction between a rotating magnetized body and a surrounding plasma (the body has, of course, a gravitational field). The

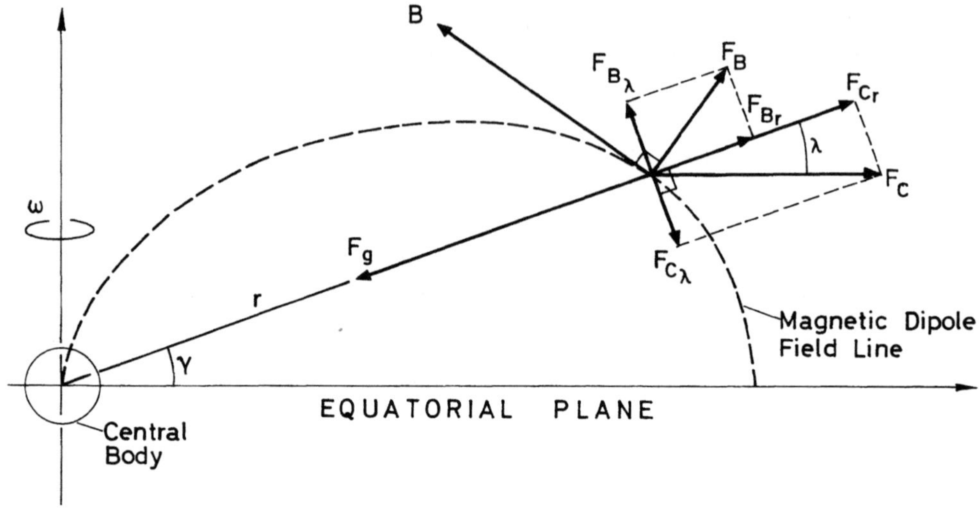

Fig. III.5. If a double layer is established a rotating central body does not bring a surrounding plasma cloud into complete corotation. The cloud instead reaches a free-wheeling state, which is determined by the equilibrium between the centrifugal force F_c, the gravitation F_g, and the electromagnetic force F_B. Under free-wheeling conditions, F_c is always smaller than F_g, and the plasma does not 'inflate' the magnetic field.

simplest case is when the magnetization is homogeneous and parallel (or antiparallel) to the rotational axis (see Figure III.5).

In the idealized case when only hydromagnetic effects are active, the plasma will attain a Ferraro corotation (CE 3.11.2). However, this is unrealistic for a number of reasons. There is always an outer limit to the corotation because of the limitations of the surrounding plasma. In the case of the Earth, the synchronous radius is comparable to the limits of the magnetosphere, outside of which we cannot expect corotation. Even closer to the dipole, there is a general sunward plasma drift, showing that the corotation is restricted to the plasmasphere. Further out the difference in angular velocity necessarily produces an e.m.f. (which tends to establish corotation) but this is taken up by electrostatic double layers. Hence the double layers produce *a decoupling* which allows a violation of the Ferraro theorem.

In the solar magnetosphere (= heliosphere or interplanetary space) the solar wind prevents the establishment of a corotation. Instead, a thin current sheet in the equatorial plane is generated. The heliospheric current system will be discussed in III.4.

The knowledge we can gain from studying the transfer of angular momentum to a surrounding plasma should serve as a basis for the approach to the early solar system problem of how the sun transferred angular momentum to the planets when they were formed and how the planets did the same to their satellite systems. The main difference from the present state is that the plasma density in the areas surrounding the central body was likely to be much larger. In the present state, the plasma in the Jovian and Saturnian magnetospheres is so thin that it can be brought into Ferraro corotation very far out. Contrary to this, in cosmogonic times, the greater density would require much larger currents to establish Ferraro corotation. But these currents would have to have been so large that they were able to produce double layers, which decoupled the surrounding plasma to a certain degree. The state which we should expect ('partial corotation' of 'free-wheeling plasma') is given by an equilibrium between the gravitation and centrifugal force which, because of the geometry of the magnetic field, should give the plasma an angular velocity of $(2/3)^{1/2}$ of the Kepler velocity (Alfvén and Arrhenius, 1976, henceforth referred to as ESS, Chapter 17). The structures of the asteroid belt and of the Saturnian rings are explained by this phenomenon.

Similar phenomena may be decisive for the evolution of galaxies. The angular momentum of the interstellar medium may be transferred to the intergalactic medium, and its kinetic energy transmitted in plasma cables to its surroundings (III.4.4).

III.4. The Heliospheric Current System

III.4.1. THE 'SECTOR STRUCTURE' AND THE EQUATORIAL CURRENT LAYER

Spacecraft measurements, which so far have been restricted to the neighborhood of the ecliptic plane, have revealed that the interplanetary magnetic field is directed inwards in certain regions, and outwards in others (see Figure III.6). The regions are separated by very sharp boundaries, obviously current layers. Their thickness is often only 1–10 proton Larmor radii (or 10^6 m) (Burlaga *et al.*, 1977). It was first thought that these current layers were perpendicular to the ecliptic plane and gave a 'sector structure' to both interplanetary space and the solar atmosphere. From a current point of view, this interpretation looked absurd because it implied that, for example, at the earth's distance, the large currents in the boundary layers should be connected along meridional planes to the sun. It seemed more likely that the observed current layers should be due to the theoretically expected equatorial current layer of Figure III.7, and that the change in field direction should be due to the position of the Earth in relation to the current layer.

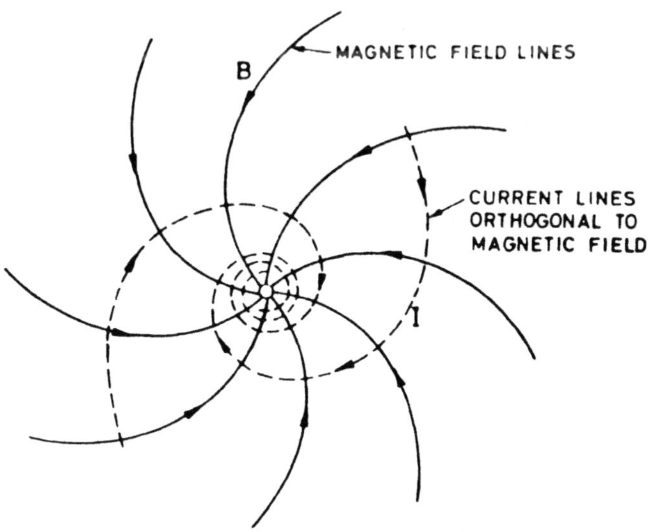

Fig. III.6. Magnetic field lines somewhat above the equatorial plane and current lines in the equatorial plane. Close below the equatorial plane the field lines have the same geometry but opposite direction.

Rosenberg and Coleman (1969) have given convincing observational evidence for the view that this theoretical picture is correct (see also, Rosenberg, 1970; Rosenberg *et al.*, 1977). These conclusions were definitely confirmed by space measurements by Smith *et al.* (1977). The phenomenon which is sometimes still erroneously referred to as the 'sector structure' of the solar wind is due to relatively small up and down displacements (10°–20°) of the solar equatorial current layer. It is similar to the wave motion of the skirt of a spinning ballerina. In our model we neglect the ballerina effect and approximate the magnetic discontinuity as a thin surface in the equatorial plane. To the north of this plane, the magnetic field lines are inward directed spirals; to the south, they are similar spirals, but directed outwards. Taking the curl of this magnetic field, we find that in the equatorial plane, there is a surface current flowing inward in a logarithmic spiral (see Figure III.6).

It seems that the direction of the magnetic field and the currents reverse when we pass from one solar cycle to the next.

III.4.2. CONSTRUCTION OF THE HELIOSPHERIC CURRENT MODEL

So far, we have done nothing more than give a formal transformation from a magnetic field description to a current description. But as soon as we speak of currents, we ask a number of questions: How does the current close? Where is the e.m.f.? Does the current transfer energy from one region to another? The answers are straightforward.

The current has a radial component, and hence, carries charge in towards the Sun. According to Kirchhoff's law, the same current must flow out from the sun and this must occur in a region which has not as yet been accessible to spacecraft measurements; i.e., at high latitudes. The simplest representation is by two currents along the axes (or in the regions near the axes). As the radial current in the equatorial plane is about 3×10^9 A (3×10^8 emu) (derived from magnetic field measurements), the axial currents must also sum to 3×10^9 A and in case of symmetry, 1.5×10^9 A (1.5×10^8 emu) each. Hence, the *translation alone* has allowed us to make the a priori unexpected prediction of the existence of currents in the axial region.

This leads to the construction of a model of the heliospheric circuit (see Figure III.7) which in its simplest possible version consists of a central body magnetized by an internal current (as in ACI), a radial current I_0 flowing in the equatorial plane and extending uniformly over 2π in longitude. Further, two axial currents each $\frac{1}{2}I_0$, must flow in opposite directions. The total current I closes at 'infinity'. The e.m.f. is due to the rotation of the central body (unipolar induction).

In a cylindrical coordinate system (r, ϕ, z), the magnetic field from I_0 is

$$B_\phi = \frac{\mu_0 I_0}{4\pi r} \quad \text{for } z > 0, \tag{25a}$$

$$B_\phi = -\frac{\mu_0 I_0}{4\pi r} \quad \text{for } z < 0. \tag{25b}$$

In order to give the magnetic field lines close to the equatorial plane the spiral shape created by the solar wind, we add a system of circular currents I_ϕ to I_0.

If the solar wind flows radially with the constant velocity v_s, and the magnetic field is parallel to the solar wind as seen from a coordinate system which rotates with the

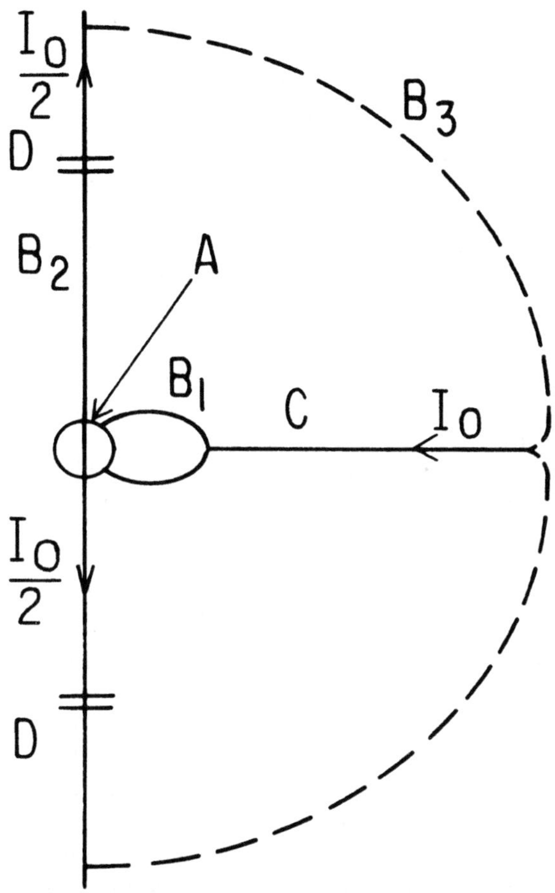

Fig. III.7. Heliospheric circuit. The Sun acts as a unipolar inductor (A) producing a current which goes outward along both the axes (B_2) and inward in the equatorial plane C_1 and along the magnetic field lines B_1. The current closes at large distance (B_3).

solar angular velocity Ω, we have $v_\phi = r\Omega$. With this idea and Equations (25a) and (25b), the magnetic field and the surface current density i become:

$$B_r = B_0(r_0/r)^2, \tag{26}$$

$$B_\phi = B_r(v_\phi/v_s) = B_0(r_0/r), \tag{27}$$

$$i_r = \frac{I_0}{2\pi r_0}\frac{r_0}{r} = i_0(r_0/r), \tag{28}$$

$$i_\phi = i_r(v_s/v_\phi) = i_0(r_0/r)^2, \tag{29}$$

where $r_0 = v_s/\Omega$ is the solar distance where the field spiral makes a 45° angle with the vector radius, B_0 is the radial magnetic field at r_0, and $i_0 = I_0/2\pi r_0$ is the radial current density at this distance. Further, we have $I_0 = 2\pi r_0 i_0 = (4\pi/\mu_0)B_0 r_0$ and with $B_0 = 2 \times 10^{-9}$ (2×10^{-5} G), $r_0 = 1.5 \times 10^{11}$ m, we find $I_0 = 3 \times 10^9$ A (3×10^8 emu).

III.4.3. PROPERTIES OF THE HELIOSPHERIC CIRCUIT

The model described above has the following characteristics, which should be compared with observed properties of the Sun and the heliosphere.

The central body acts as a unipolar inductor and the e.m.f. is produced in region A (see Figure III.7). The mechanical force on the solar atmosphere $d\mathbf{F} = I\, ds \times \mathbf{B}$ tends to decelerate the rotation of the central body. The current transfers angular momentum from the central body to the surrounding plasma. Hence, we have a decelerating force applied to the solar atmosphere in the polar region. This should produce a *non-uniform rotation* of the Sun of the type which is observed (angular velocity decreasing with increasing latitude. Whether this interpretation is the correct quantitative explanation of the non-uniform rotation is an open question.

In region B_1, the currents are field-aligned. It seems to be a general rule of cosmic physics that field-aligned currents frequently manifest themselves as luminous filaments (II.4). If the current in B_1 is spread over an extended region, we should expect filaments. *Equatorial streamers* in the solar corona may be explained in this way.

Similarly, in the polar region, the vertical currents near the solar surface may produce the *polar plumes* in the solar corona.

The model predicts that there should be currents near the axis strong enough to match the current in the equatorial plane. Such currents should be observed when a spacecraft is sent to the high latitude regions. It is an open question to what extent they flow *very* close to the axis. They may be distributed over a large region and may in part flow at medium latitudes.

Electrostatic double layers (and/or mirror produced voltage drops) may be formed, especially by the axial currents. This means that we may have regions D far away from the Sun, where energy is released without any observable indication of how it is transferred. This is analogous to the release of energy in the double layers of the auroral zones. If a double layer is formed far out from the Sun, it may emit radio waves or plasma waves which could be detected. A possible application of this is discussed in the next section.

III.4.4. EXTRAPOLATION TO GALACTIC DIMENSIONS: DOUBLE RADIO SOURCES

If in the heliospheric circuit we replace the rotating magnetized sun by a galaxy, which is also magnetized and rotating, we should expect a similar current system, but magnified by about 9 orders of magnitude. This seems to be a very large extrapolation, but in fact the successful extrapolation from the laboratory to the magnetosphere is by almost the same ratio. (Of course all theories of plasma phenomena in regions which cannot be investigated by *in situ* measurements are speculative!)

The e.m.f. is given by Equation (20), taken from the galactic center out to a distance where the current leaves the galaxy, which may be the outer edge. Inside the galaxy the current may flow in the plane of symmetry similar to the current sheet in the equatorial plane of the Sun, but whether the intragalactic picture is correct or not is not really important to our discussion here. The e.m.f. which derives from the galactic rotation is applied to two circuits in parallel, one to the 'north' and one to the 'south' (see Figure III.8). As galaxies in general are highly north-south symmetric it is reasonable that the

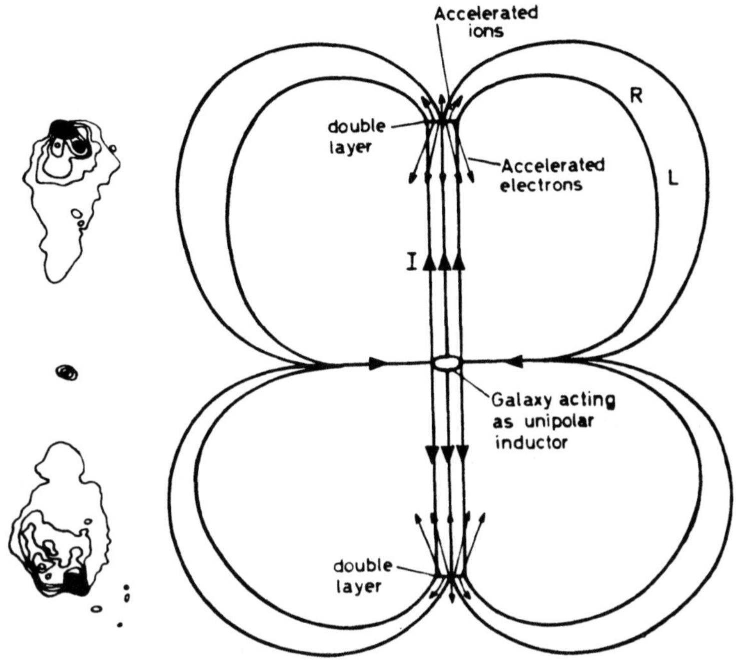

Fig. III.8. Galactic circuit. (a) The heliospheric circuit is scaled up by a factor 10^9 and the Sun is replaced by a galaxy. (b) Observed radio emission of Cygnus A (by Hargrave and Ryle 1974) is attributed to synchrotron emission by electrons accelerated in the double layer. The galaxy delivering the energy is located almost exactly between the radio sources.

two circuits are similar. Hence, we expect a high degree of symmetry in the current system (at least under idealized conditions).

In the magnetosphere, the current flowing out from the ionosphere produces double layers (or magnetic mirror induced fields) at some distance from the Earth (II.5.2 and III.2). Because of the similarity of the plasma configuration, we may expect *double layers at the axis of a galaxy*, and a large release of energy in them. It is suggested that *the occurrence of such double layers is the basic phenomenon producing the double radio sources*. This agrees with Hargrave and Ryle's (1974) conclusion that energy must be continuously supplied to the radio sources from the central galaxy.

In the galactic circuit, the e.m.f. is produced by the rotating magnetized galaxy which implies that the energy is drained from the galactic rotation. By the same mechanisms as in the auroral circuit, it is transferred to the double layers where the power $P = IV_D$ is released. In one or more double layers on each side of the galaxy, an acceleration of charged particles takes place. From the magnetosphere, we know that layers are produced when the current flows outwards (whether double layers can be formed when the current flows inwards is still an open question). If the same is true in the galactic case, there is a flow of thermal electrons to the layer from the outside and when passing a series of double layers the electrons are accelerated to very high energies. Hence, *a beam of very high energy electrons is emitted from the double layer along the axis towards the central galaxy*. This process is the same as the one which produces auroral electrons, only scaled up enormously both in size and energy. In analogy with the current in the magnetotail (III.5), the current in the equatorial plane of a galaxy may also produce double layers, which may be associated with large releases of energy.

Figure III.8 shows a radio astronomy picture of a double radio source. It is essential in our model that the e.m.f. of the galaxy has such a direction that the axial currents flow outwards. The double layer they produce should be located at the outer edges of the strong radio source. When electrons conducting the currents outside the double layer reach the double layer, they are accelerated to very high energies. Similarly, ions reaching the double layer on their outward motion from the central galaxy, will be accelerated outwards when passing the double layers. The strong axial current produces a magnetic field, which pinches the plasma, confining it to a cylinder close to the axis.

Although the electrons are primarily accelerated in the direction of the magnetic field, they will be scattered by magnetic inhomogeneities and spiral in such a way that they emit synchrotron radiation. The accelerated electrons will be more like an extremely hot gas than a beam. With increasing distance from the double layer the electrons will spread and their energy and hence their synchrotron emission will decrease. This is in agreement with observations. It is possible that some of them will reach the central galaxy and produce radio emission there. It is also possible that the observed radio emission from the central galaxy is due to some other effect produced by the current (there are several mechanisms possible). Such phenomena in the central galaxy will not be discussed here.

The ions passing the double layer in the outward direction will be accelerated to the same energy as the electrons. Because of their larger rest mass, they will not emit much synchrotron radiation, but there are a number of other mechanisms by which they may produce the observed radio emission from the regions farther away from the central galaxy.

It should be stressed again that, just as in the magnetosphere and in the laboratory, *the energy released in the double layer is transferred to it by electric currents which essentially consist of relatively low energy particles*. There is no need for a beam of high energy particles to be shot out from the central galaxy (and still less for some mysterious 'plasmons'). On the contrary, the central galaxy may be bombarded by high energy electrons which have obtained their energy from the double layer. For the details of the theory and comparison with observations, see Alfvén (1977).

The quantitative analysis of the double radio galaxies shows that the plasma current must be of the order $I = 3 \times 10^{17}$ A, while the voltage drop in one, or more likely, a series of double layers, should be of the order of $V_D = 3 \times 10^{16}$ V. This brings us up to the range of the most energetic cosmic radiation. In fact the double layers should emit ions with an energy

$$W = ZeV_D \tag{30}$$

where Z is the charge of the ion. With V_D given above, an ion with $Z = 30$ receives an energy of 10^{18} eV.

Also, betatron acceleration (CE 2.3.1) in a pinching current may give energies in the same range. In fact, if a current I produces a magnetic field $B = \mu_0 I/2\pi R$ this field can contain and accelerate particles with a rigidity *Br*. The total energy delivered as very high energy cosmic rays should be an appreciable fraction of the energy of the synchrotron radiation from the radio source.

It need not be stressed that the model suggested here is necessarily speculative, like all models for regions outside the reach of spacecraft. See Figure I.2.

III.5. Circuits of the Magnetospheric Tail, Comets, and Venus

III.5.1. TAIL CIRCUIT AND MAGNETIC SUBSTORMS

The circuit of the current which magnetizes the magnetotail is in principle very simple (Figure III.9). It consists of a sheet in which a current flows from a to c in a plasma which, in relation to the Earth, is at rest or moves slowly. The circuit is closed by currents $cb'a$ and $cb''a$ flowing in the solar wind, which has velocity v perpendicular to the plane of the figure and a magnetic field B. The e.m.f. is

$$V = \int_{cb'a} (\mathbf{v} \times \mathbf{B})\, \mathrm{ds} \tag{31}$$

and when B has a southward component close to the magnetotail it produces a current system as shown in the figure. This circuit is part of the large scale magnetosphere circuit (III.6).

In the sheet ac, an electric double layer may be formed, and it frequently explodes, so that the current ac is disrupted. However, because of the large inductance of the whole circuit, the electric current cannot be stopped immediately. According to Boström (1974), the result is that the current chooses a new path: it flows along the magnetic

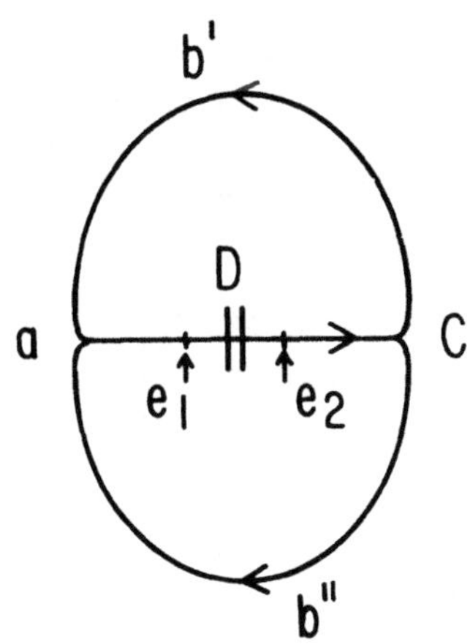

Fig. III.9. Tail circuit. The line elements $cb'a$ and $ab''c$ are located in the solar wind, and aDc in the magnetotail. The polarization of the solar wind produces an emf of $\int_{cb'a} \mathbf{v} \times \mathbf{B}\, \mathrm{ds}$ which gives a current in the neutral sheet in the tail. This current system determines the magnetic field in the tail. An electric double layer may be formed at D, which, if it is exploding, produces a magnetic substorm, which may be repetitive according to Figure II.21.

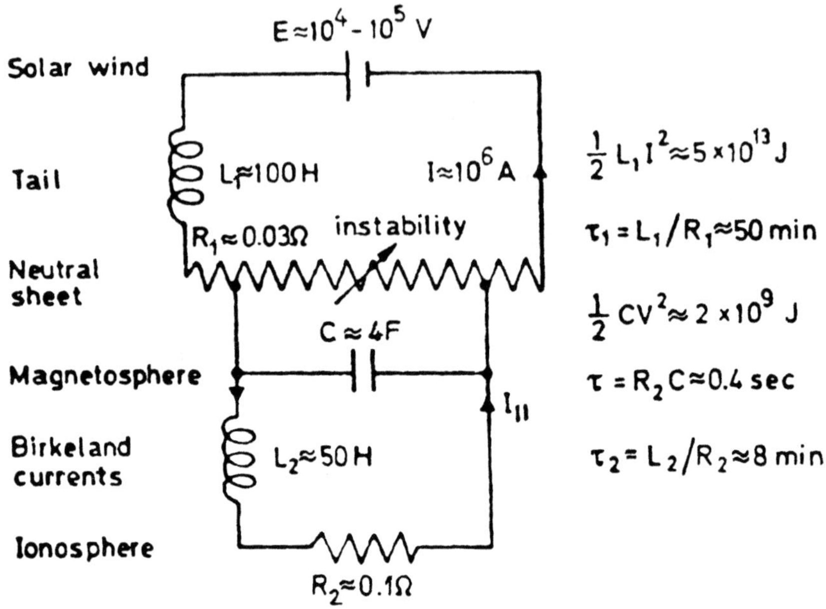

Fig. III.10. Equivalent magnetic substorm circuit (after Boström, 1974). At substorm onset, the resistance of the neutral sheet increases and the tail current is redirected to the ionosphere.

field lines to the ionosphere. This seems to be a reasonable explanation of *magnetic substorms*. The phenomenon can be described by drawing the circuit shown in Figure III.10.

Investigations show that a number of different phenomena are closely related with magnetic substorms. Movements of the dayside cusp (see Fig. I.2(c)) and the associated dayside aurora show a dependence on substorm activity and changes in the direction of the interplanetary magnetic field (Mende *et al.*, 1980). Such apparent correlation indicates the complicated nature of the mechanisms involved.

III.5.2. CURRENT SYSTEM IN COMETS

With the magnetotail circuit as a model Mendis, partially in collaboration with Ip, has constructed the current system of comet tails (see Mendis, 1978). A current is supposed to flow in a plane of symmetry in the tail so that the magnetic field indicated by the filamentary structure (see Figure II.6) goes in opposite directions in the two halves of the tail. The current in the plane of symmetry may be disrupted in the same way as in the terrestrial magnetotail. The 'folding umbrella' phenomenon, which is a sudden change in the angle between the filaments in the tail, should be caused by such a current disruption. This is similar to the events leading to a terrestrial substorm.

III.5.3. CURRENT SYSTEM IN THE MAGNETOSPHERE OF VENUS

As Venus, like the comets, has no appreciable intrinsic magnetic field, the solar wind interaction with her is likely to be essentially the same. Gringauz (1980) has analysed the *in situ* measurements made by a Venera mission to Venus and constructed a current

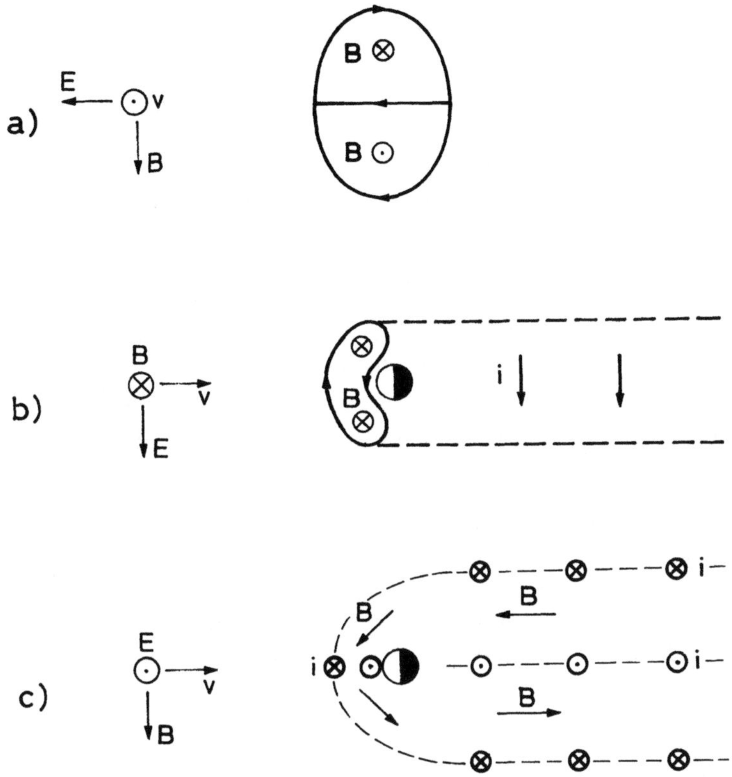

Fig. III.11. Current system of Venus. (a) shows the tail circuit as seen looking towards the sun. The current system is depicted in (b) as seen from the north and in (c) from the Cytherean dawn side (Gringauz, 1980).

system which is similar to the Mendis model, which hence to a certain extent has been confirmed by *in situ* measurements.

The Gringauz current (see Figure III.11) system has a plane of symmetry which is parallel to the solar wind electric field (i.e., perpendicular to the solar wind magnetic field component perpendicular to the vector radius). Hence the plane of symmetry changes when the solar wind magnetization changes direction. In the tail of Venus, i.e., in its shadow, there is a *neutral sheet* with $B = 0$ in the plane of symmetry. A surface current flows in this sheet and closes through semi-circular loops 'upwards' and 'downwards' from the plane of symmetry. Hence the tail circuit is in principle the same as the tail circuit in Figure III.9.

At the front side there is a current which is essentially the same as in the model (c) in III.1.3, where the conductor is represented by the Cytherean ionosphere. The measurements by Elphic *et al.* (1979) described in II.4.7 refer to this ionospheric current. Both the tail and the front current systems are measured directly. It is likely that they are combined as shown in the figure, but this is an extrapolation. Compare also Figure II.2 (neglecting the auroral current system).

The described current system gives a magnetic field which can be depicted as solar wind field lines wrapped up by the obstacle which it meets.

The main difference between Venus and a comet is that the gravitation of Venus prevents gases from its ionosphere to evaporate into its tail so that it is not visible like the tail of a comet.

The similarity between the current systems of Venus and comets and the terrestrial magnetotail makes it likely that there should be substorms or folding umbrellas at Venus.

III.6. Magnetospheric Circuit

The magnetic field of the Earth makes its current system more complicated than those of non-magnetized bodies. Starting from the Mendis–Gringauz model we can say that – summarized in the most simple way – the intrinsic magnetic field has the following effects. It splits the current system into a front system and a tail system and inserts the auroral system between these.

The terrestrial correspondence to the current in the upper ionosphere of Venus is repelled far out from the atmosphere and instead flows in the magnetopause. The flux through the front circuit encloses the Earth's dipole field which is greatly distorted. The flux in the tail derives partly from this flux, but mostly from the dipole field. The result is a very complicated geometry. (Models of the current system of the magnetosphere have been constructed by several authors. One of the earliest attempts is due to Cole (1974, 1976); see also Potemra (1979).

We shall approach the problem by successive approximations (Alfvén, 1979). We start with the simplest possible model, viz., a magnetic dipole field ($\mu_{\oplus}$ = dipole moment) representing the terrestrial field, on which is superimposed a homogeneous interplanetary field B_s, either parallel or anti-parallel to the dipole axis, and a uniform solar wind with velocity v_s perpendicular to the dipole axis. The magnetic model is immersed in an electric field which at large distances from the Earth has the constant value

$$\mathbf{E}_s = \mathbf{v} \times \mathbf{B}_s. \tag{32}$$

III.6.1. ZERO ORDER APPROXIMATION: ONE-PARTICLE PROBLEM

Our first case is when the plasma flux of the solar wind is so close to zero, that the Debye length is long compared to the size of the magnetic configuration. In this case, the motion is a one-particle problem (Alfvén, 1939, 1955). The interplanetary magnetic field must be northward in order to allow the particles to penetrate into the magnetosphere.

III.6.2. FIRST ORDER APPROXIMATION: PLASMA FLUX SMALL

In the next case, we still assume the plasma flux to be so small that the magnetic perturbations are negligible, but this time, the density is considered large enough for the Debye length to be small compared to a characteristic length. Hence, space charge is important. Figure III.12 shows the field in case the interplanetary magnetic field is directed northward. The magnetic field line which at infinity coincides with the dipole axis, is split and surrounds the entire magnetosphere like a cocoon. As there is a free flow of charge along these field lines, there can be no voltage difference in the cocoon on which a space charge is present. In this way the magnetosphere is screened from the solar wind.

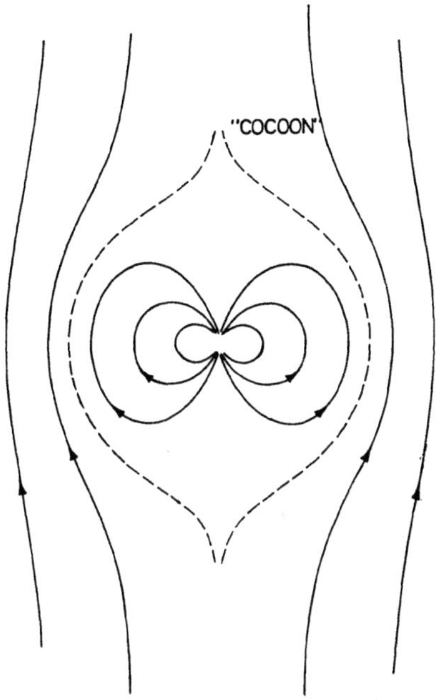

Fig. III.12. Earth dipole field combined with northward interplanetary field. Coupling between the dipole field and solar wind field is a minimum.

If instead the interplanetary magnetic field is southward, a strong coupling between the interplanetary field and the magnetosphere is produced (see Figure III.13).

This configuration has the following properties:

(a) A neutral line ($B = 0$) is produced in the equatorial plane ($z = 0$) at a distance r_n from the dipole, with

$$r_n = \left(\frac{\mu_0 \mu_\oplus}{4\pi B_s}\right)^{1/3}. \tag{33}$$

This is defined as the limit of the magnetosphere.

(b) For $z \to \infty$, the field lines through the neutral line are located at a circle

$$r_s = r_n \sqrt{3}. \tag{34}$$

(c) The homogeneous electric field E_s at infinity causes a voltage difference

$$V = 2r_s E_s \tag{35}$$

between a and d and between a' and d' (see Figure III.14). The magnetic field configuration transfers this to the magnetosphere.

This configuration should be compared to model (a) in III.1.3 (Figure III.1(a)) in which a plasma enters a deceleration region where inertia currents tap part of the plasma kinetic energy into (inductive) circuit energy. In our case, solar wind energy is tapped by inertia currents ad and $a'd'$, and transferred to the magnetosphere by the same mechanism.

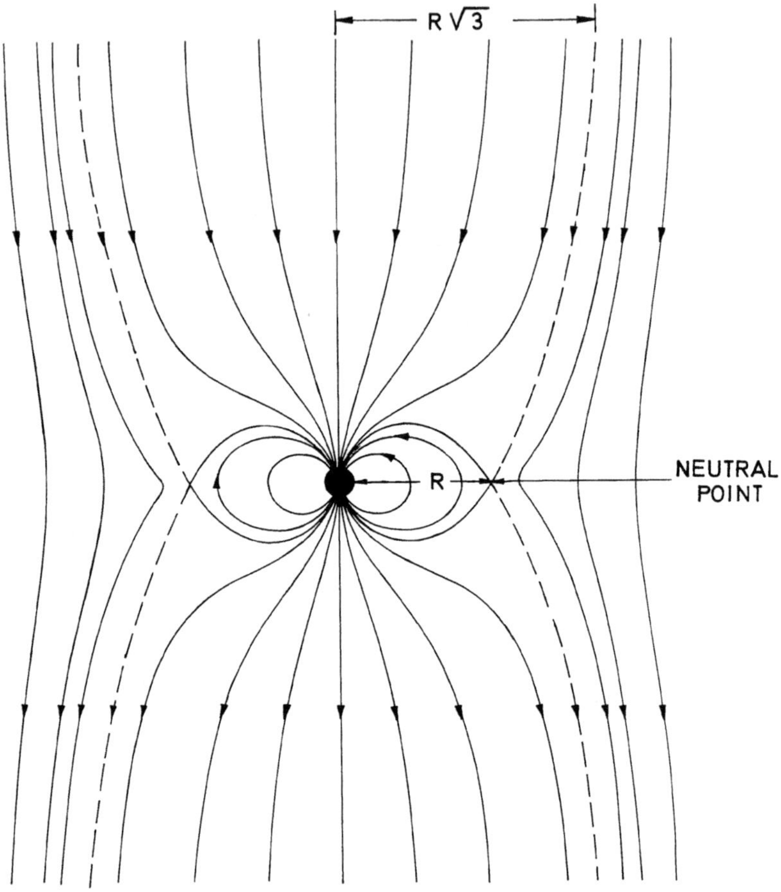

Fig. III.13. Earth dipole field combined with southward interplanetary field. Coupling between the fields is a maximum.

Thus, we have a *transfer of energy from the solar wind to the magnetosphere* (Figures III.14(a, b)). The value of R_D in Fig. III.1(a) should be calculated from the losses.

The coupling between the solar wind and the magnetosphere has two main effects:

(a) The electric field which is transferred along the magnetic field lines produces a sunward plasma drift in the equatorial region of the magnetosphere. In order to produce and sustain this, a certain tapping of solar wind energy is necessary.

(b) It produces currents along the sunward and the antisunward sectors of the neutral line. As the field lines which go directly from the polar ionospheres to infinity have very large mirror ratios, currents along these may not be of primary importance. Hence, we should expect the currents to flow mainly along the lines of force connecting the neutral line with infinity. Since the magnetosphere acts as an impedance in the circuit, the inertia currents *ad* and *a′d′* will produce a voltage drop along these lines and therefore decelerate the solar wind.

The inertia currents close the circuit as is shown in Figures III.14 and III.15.

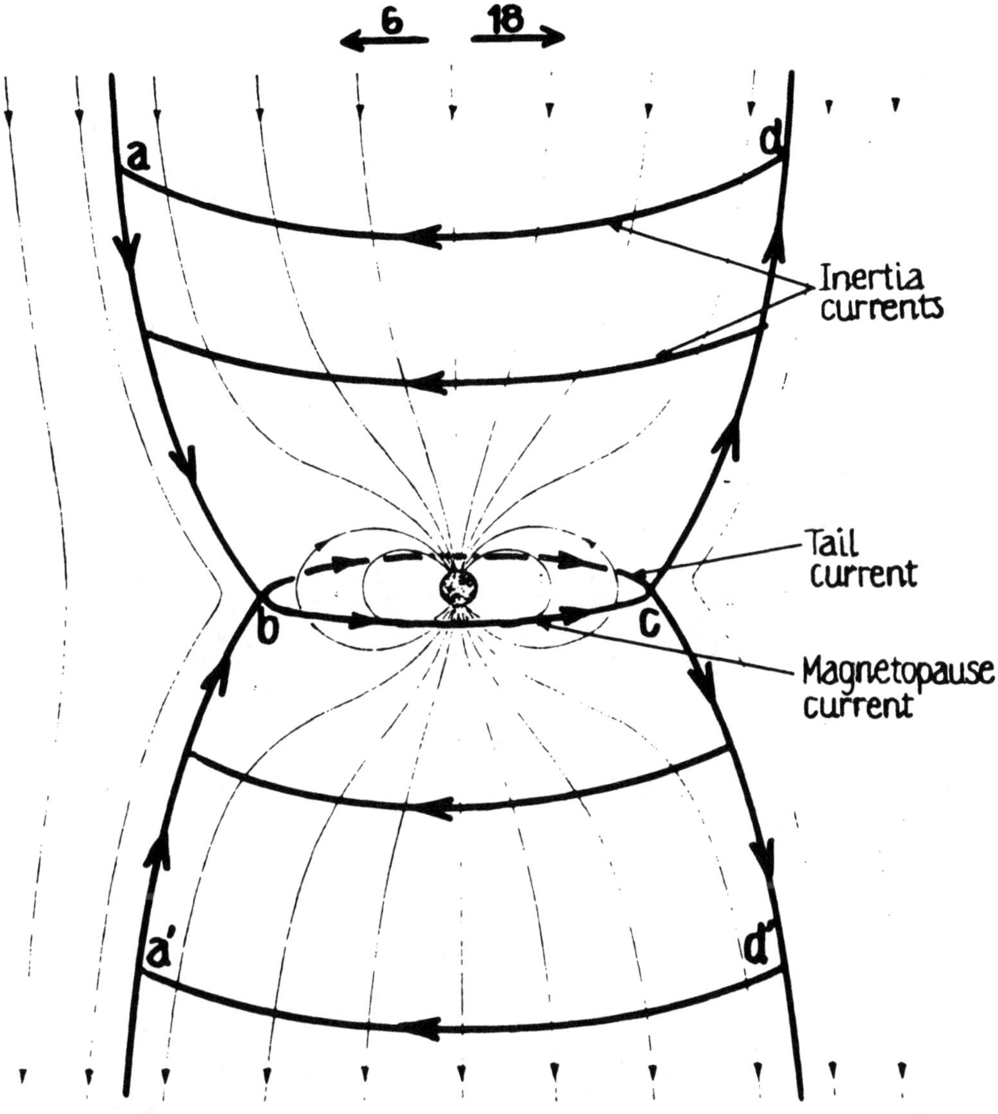

Fig. III.14(a). Current system seen from the Sun. The solar wind electric field produces a voltage difference between the field lines *aba'* and *dcd'*, which produces a current along the sunward and the antisolar parts of the neutral line *bc*. It also produces field-aligned and perpendicular currents over the magnetosphere.

III.6.3. THE THREE FIRST-APPROXIMATION CIRCUITS

Depending on how the currents flow in the magnetosphere, we can define three different *circuits for the transfer of solar wind kinetic energy to the magnetosphere.* (There are also electric currents in the solar wind which do not flow to the magnetosphere.) In all three cases, solar wind energy is tapped through inertia currents.

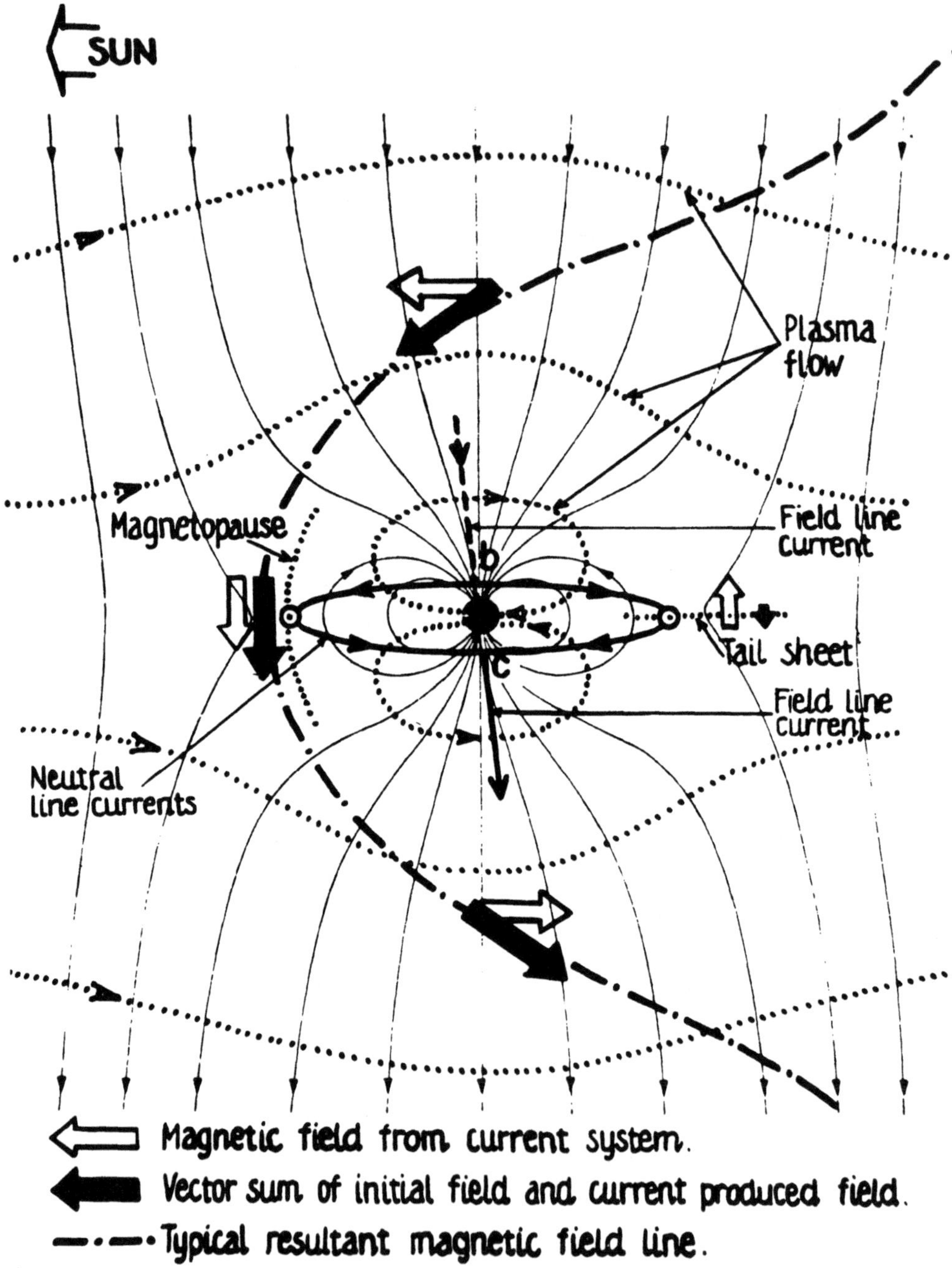

Fig. III.14(b). Current system seen from dusk side. This also shows the plasma flow and the resultant magnetic field configuration. The field-aligned currents to the dawn and dusk sides of the neutral line are closed through the sunward part of the neutral line (resulting in the formation of the magnetopause) and the antisolar part (resulting in the tail neutral sheet).

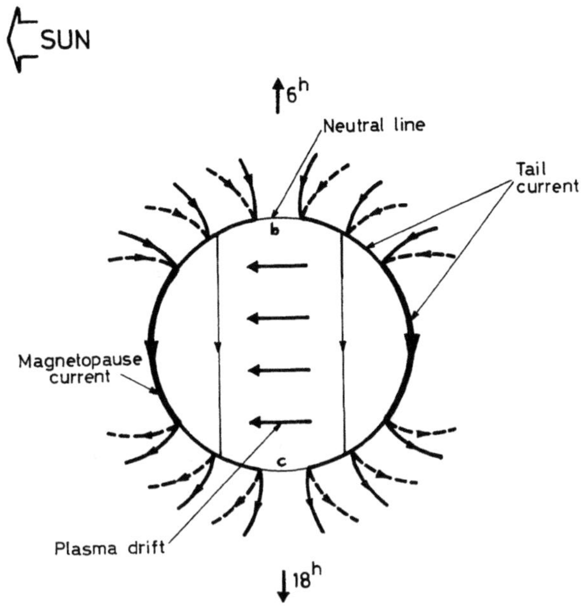

Fig. III. 15. Solar wind–auroral circuit. It shows the neutral line from the north.

A. *The magnetopause circuit.* The current closes along the sunward part of the neutral line *bc* in Figures III.14 and III.15 (for the motivation of the name, see second approximation, III.6.5).

B. *The tail circuit.* The current closes through the tail part of the neutral line *bc* (Figures III.14 and III.15).

C. *Solar wind–aurora circuit.* This circuit produces and sustains the sunward plasma drift in the equatorial region of the magnetosphere. As this flow is likely to produce discharges over the auroral region, a large part of the power in this circuit is dissipated as aurora (hence the name of the circuit).

III.6.4. PHENOMENA PRODUCED BY FIRST APPROXIMATION CURRENTS

As a summary, the first order current system produces the following effects.

A. *Changes of the Large Scale Magnetic Field*

It is easily seen that the current system produces changes in the large scale magnetic field of the type illustrated in Fig. III.14(b).

B. *Magnetospheric Drift*

The transfer of the interplanetary electric field to the magnetosphere produces a sunward drift in the region close to the equatorial plane of the type depicted in Figure III.15 (the influence of the plasmasphere is neglected). The electric field associated with the drift compensates for the applied electric field. Magnetic field lines from the ionosphere may partially shortcircuit the voltage, so that currents from the equatorial plane flow to the auroral zone and back again both in the 6^h region and in the 18^h region, in agreement with the Armstrong–Zmuda current system (Armstrong, 1974; Zmuda and Armstrong, 1974). The circuit is closed by ionospheric currents, especially in the auroral zone.

III.6.5. SECOND ORDER APPROXIMATION. REALISTIC PLASMA FLOW

We shall now see what happens if we increase the very small solar wind plasma flow to realistic values. Obviously, the magnetosphere, including the current system, will be considerably deformed. However, the topology will probably not change in a decisive way. It seems reasonable that, as the next step we scale up the perturbations of the first approximation. The purpose of this second order approximation, the characteristics of which are related in III.6.6–III.6.10 and III.7, is to serve as a basis for a third order approximation, which will not be attempted in this monograph.

III.6.6. MAGNETIC FIELD CHANGES

The large scale magnetic field should be the vector sum of the primary magnetic field and the perturbation field. Figure III.14(b) shows the result. It is obvious that a comet-like magnetic field is obtained, which is rather similar to the observed field. The neutral line currents will be transformed into sheet currents by the Dungey effect (Dungey, 1961, 1964). On the sunward side of the magnetosphere, the line current will spread perpendicular to the equatorial plane, and a sheet will be formed carrying a strong current. We identify this with the *magnetopause*. It is energized through the magnetopause circuit. In a related way, the line current on the back side will spread into a sheet current, but this will be located in the equatorial plane. This should be identified with the *tail neutral sheet*.

III.6.7. FRONT (SHOCK FRONT) CIRCUIT

In the first approximation, the solar wind near the equatorial plane will experience a decreasing magnetic field when it approaches the neutral line. When the magnetopause develops, the magnetic field in front of it will increase. The result is a slowing down of the solar wind drift, which is accompanied by an inertia current opposite to the magnetopause current. The situation is analogous to the case treated in model (c) of III.1.3. A plasma approaching an obstacle (in this case, the magnetopause) is stopped and deviated, not directly by the obstacle, but by the electric field it generates in the approaching plasma. As long as the magnetic field perturbation deriving from this inertia drift is negligible, the current is distributed over a layer of considerable thickness (Alfvén, 1955; because of the reversed interplanetary field the sign of the current in this paper should be reversed). When the plasma density is so large that the inertia current greatly perturbs the magnetic field, the current layer is contracted to a thin sheet. This is identical with the 'shock front' derived by the hydromagnetic approach. Our model predicts that if the solar wind flow is very low, there should be a distributed current instead of a shock.

In some hydromagnetic pictures of the 'shock front', it is not observed that the current associated with the 'shock' necessarily must close in some way. In the particle description, this is obvious (see model (c)). The force $\mathbf{i} \times \mathbf{B}$ in the front layer decelerates the plasma and brings it almost to rest in front of the magnetopause. It must then be accelerated sideways. As in the particle description of model (c)), this can only be done by electromagnetic forces. We conclude that there must be currents between the front layer and the magnetopause which produce such an acceleration by the $\mathbf{i} \times \mathbf{B}$ force.

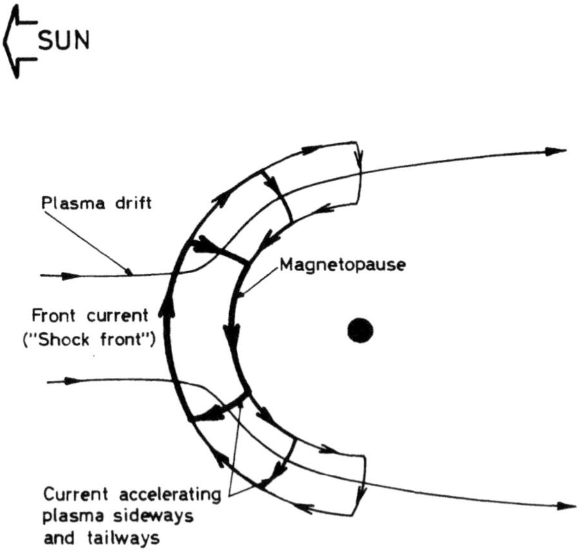

Fig. III.16. Simple picture of the front (shock front) current circuit outside the magnetosphere of the Earth (compare Figure III.1(c)). The front current decelerates the solar wind in front of the magnetosphere. Currents connecting the front current with the magnetosphere accelerate the plasma sideways and tailways. The currents close through the magnetopause. (In the coronal current loops the situation is expected to be the reverse with the currents accelerating the solar wind plasma.)

Hence, we derive the circuit of Figure III.16. The front current closes through a current in the magnetopause, which is added to the current deriving from the magnetopause circuit. This is again analogous to model (c). Compare also the front circuit in the magnetospheres of Venus and comets (Figure III.11).

III.6.8. MAGNETOPAUSE CIRCUIT

The magnetopause circuit is basically the same as in the first order approximation; it transfers energy from the solar wind to the magnetopause. However, the neutral line current in the first approximation is now changed into a surface current, and it is not quite obvious how this connects with the solar wind currents. Furthermore, as part of the magnetopause current is now furnished by the front circuit, there is a strong coupling with this which may affect its behaviour.

III.6.9. SOLAR WIND – AURORAL CIRCUIT

In the region between the magnetopause and the tail sheet, the Sunward drift of the magnetospheric plasma will again be basically unchanged from the first approximation. As soon as the currents to the auroral ionosphere exceed a certain value, double layers will be formed, but only by currents going upwards from the ionosphere (Lennartsson, 1977). In the double layers, an acceleration of charged particles will take place, and these are the cause of at least the more brilliant auroras. Combining Figure III.15 with Figure II.17 we obtain a rather straightforward transfer of kinetic energy from the solar wind to the high energy auroral particles. In other words, the solar wind

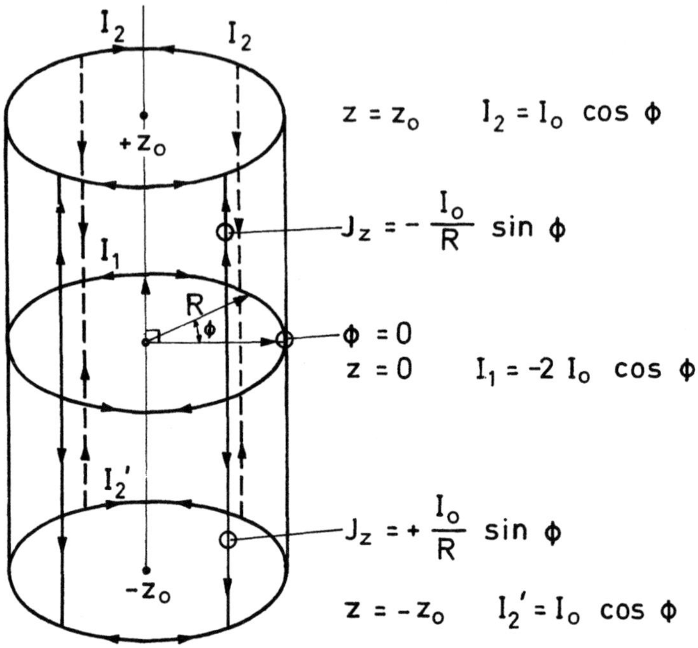

Fig. III.17. The Three-Ring Circuit Model. The center ring represents the current in the neutral line. The vertical currents in the cylinder sheet represent field-aligned currents to the neutral line. The other two ring currents above and below the equatorial plane close the current system and represent distributed perpendicular currents.

generator is connected by a high power transmission line directly to the auroral consumer of energy. For details see Alfvén (1979).

III.6.10. THE TAIL CIRCUIT

The neutral line current in the first approximation is changed into a sheet current. As with the magnetopause circuit, it is not very clear how its connections with the solar wind currents are modified.

The current flows through the neutral sheet in the tail, producing the characteristic tail structure. The circuit converts solar wind energy into inductive energy of this circuit, manifesting itself as a change in the tail magnetic field. When the neutral sheet current exceeds a critical value, an explosive double layer is formed, which is likely to be the cause of a magnetic substorm. The tail current is disrupted (at least partially) and the current is redirected over the auroral zone. This is likely to be the basic mechanism of a magnetic substorm (see Boström's model, Figure III.10). The great amount of energy released by a magnetic substorm, shows that the tail current carries an important part of the total energy delivered by the solar wind to the magnetosphere.

III.6.11. THIRD APPROXIMATION AND COMPARISON WITH OBSERVATION

When the first and second approximations are worked out in detail, they should serve as a platform for a third approximation, which hopefully should give a good picture of observational reality.

It is not certain whether the solar wind–aurora circuit is the most important source of energy for the aurora. It is quite possible that a large, perhaps the largest, source of energy for the aurora derives from the tail circuit, as suggested by Boström (1974). There is a great inflow of new observational material and the conclusions are still controversial in several respects.

III.6.12. THE THREE-RING MODEL

It is important to develop the qualitative magnetospheric model that we have discussed so far, into a quantitative model. In spite of the fact that we have chosen a highly idealized and simplified model, this task still encounters formidable mathematical difficulties. In order to make the problem tractable, we have to make further simplifications.

We substitute for the complicated geometry of Figure III.14, a cylinder having the same radius r as the neutral line. On the surface of the cylinder in Figure III.17 the following currents flow: Three circular ring currents (after which the model was named), one (I_1) at the neutral line and two (I_2 and I_2') on both sides of it at a distance $+z_0$ and $-z_0$. The line currents are

$$I_1 = -2I_0 \cos\phi, \tag{36}$$

$$I_2 = I_2' = I_0 \cos\phi. \tag{37}$$

These currents are joined by surface currents in the z direction with the density

$$J_z = \pm\frac{I_0}{R}\sin\phi \tag{38}$$

where the current parameter I_0 can be chosen arbitrarily, although we will relate its magnitude to the interplanetary electric field $\mathbf{E}_s$ through Equations (32) and (35).

The main idealizations are that the distributed inertia currents are replaced by line currents, and that the radii of these are reduced to the same size as the neutral line.

Whipple *et al.* (1980) has derived the magnetic field from a dipole field and a homogeneous southward interplanetary field (see Figure III.13), on which the field from this current system is superimposed. The distance between the neutral line current and the concentrated inertia currents is arbitrarily put equal to $70R_\oplus$. Solutions with currents $I_0 = 3.2 \times 10^5$ A and 9.6×10^5 A are shown in Figure III.18(a) and Figure III.18(b) respectively. In the first case the deformation of the field from the current system is so small that it is still reasonable to use the current model of Figure III.17 as a first approximation, but the absolute value of the current is unrealistically small. In the second case, the current value is realistic but the deformation of the field is so large that the three ring model is unrealistic.

In both cases it is unsatisfactory to use a line current along the neutral line, because the Dungey effect (1961, 1964) must change the line currents into surface currents: the surface stands perpendicular to the equatorial plane in the magnetopause region and is located in the equatorial plane in the tail. From an analytical point of view, the more realistic model with surface currents instead of line currents introduces considerable complications which have not yet been overcome.

The computer results for these two cases are shown in Figure III.18. In Figure III.18(a) it is shown that the current system has the following effects.

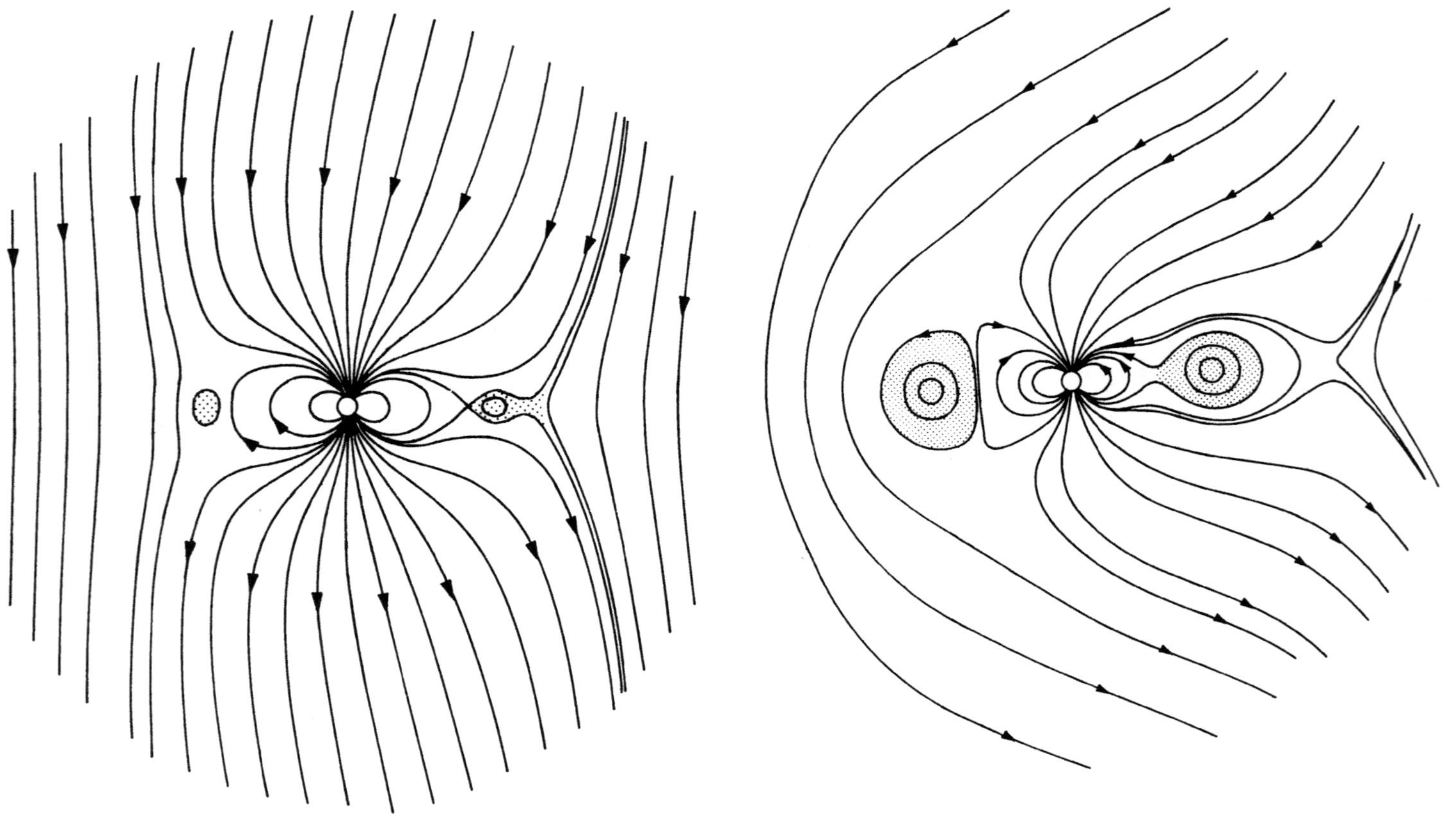

Fig. III.18. Computer results of the three ring model. We have a magnetic dipole and a homogeneous southward solar wind field, on which the field from the current system (neutral line current and inertia current) is superimposed. (a) Magnetic field configuration for a current value (I_0) of 3.2×10^5 A. The shaded areas are artifacts due to neutral lines instead of neutral surfaces representing the magnetopause and the tail neutral sheet. (b) Magnetic field configuration for current value (I_0) of 9.6×10^5 A. The Sun is to the left, with the solar wind flowing from left to right.

(a) The magnetic field lines far away from the Earth bend anti-Sunwise.

(b) Pronounced cusps are produced on the Sunward side.

(c) Singular points are produced by the unrealistic line current model, both on the sunward and anti-sunward side. No account should be taken of the computed fields very close to these points (shaded regions).

Figure III.18(b) shows the same effects, but the effects are more pronounced. The bending of the magnetic field lines is so large that the model resembles the observed field. The cusps have moved polewards. The unrealistic shaded areas which are due to the use of line currents instead of surface currents, are larger. A realistic field in these regions requires a model in which the current flows in sheets. A qualitative discussion of what changes this will introduce indicates that this may result in a field not very different from the observed field.

The results obtained with this approximate model are so encouraging that an attempt should be made to develop a third approximation model based on the Whipple model. However, this will encounter large analytical difficulties.

The conclusion we can draw is that it seems possible to derive essential features of the magnetic field configuration from a very simple and reasonably straightforward theory based on the particle description of a plasma. It is evident that the deformation of the terrestrial magnetic field is not due to the solar wind sweeping the field lines with it, but can be accounted for as a result of the current system which the solar wind produces. Hence in this case the frozen-in field line concept is appealing, but in reality pseudo-pedagogical.

III.7. Other Magnetospheres

In situ measurements have been made in the magnetospheres of Mercury, Mars, Jupiter and Saturn. Some current models have been constructed, but so far no complete system capable of accounting for the transfer of energy from the solar wind. It is not quite clear whether we have enough observational data for deriving such a system.

III.8. Solar Prominence Circuit and Solar Flares

As the Sun is still outside the reach of spacecraft, detailed diagnostics of solar plasmas is impossible, and theories of solar phenomena will necessarily remain speculative, or in any case, unconfirmed by high-quality diagnostics. However, if we combine solar observations with careful extrapolations from regions which have been explored by spacecraft, we may clarify several solar physics problems.

Because of its filamentary structure, we conclude that a solar prominence is due to a current. In the photosphere, there are motions with velocity $\mathbf{v}$, which are sometimes irregular (e.g., originating from the granulations), and sometimes systematic (e.g., from whirls around Sun-spots, or from the non-uniform rotation of the Sun). As the photosphere is magnetized (field = $\mathbf{B}$), its electric voltage V varies from point to point. If a magnetic field line above the photosphere runs between point A to another point B, the voltage, $V = \int_A^B \mathbf{v} \times \mathbf{B}\, ds$, between A and B is applied to this field line, and may cause an electric current along the field line. The circuit is similar to the auroral circuit, but simpler; it consists of a magnetic flux tube intersecting the solar surface at two points, and a photospheric or subphotospheric connection between these points.

With $v \simeq 10^4$ m s^{-1}, $B = 10^{-3}$ T (10 G), given the distance between A and B to be of the order of 10^8 m, we find $V = 10^9$ V. However, if **v** and **B** are irregular (vary in a more or less random way), V could have much smaller values. On the other hand, the values for v and B are by no means extreme. We may put $10^8 - 10^9$ V as typical values. According to Alfvén and Carlqvist (1967), the electric current is expected to be $10^{11}-10^{12}$ A ($10^{10}-10^{11}$ emu), a value which has been confirmed observationally by Moreton and Severny (1968). The inductance of the circuit is 10 H, and the stored energy is $\frac{1}{2}LI^2 = 10^{23}$ joule (10^{30} erg).

The size of a prominence circuit is comparable to that of the auroral circuit. The currents are much larger, because of the higher e.m.f. and the higher conductivity (the high resistivity of the upper ionosphere is replaced by a low resistivity due to the high temperature). As in the auroral circuit, a double layer may be produced. The auroral double layer is often rapidly fluctuating, giving rise to the rapid variability of auroral phenomena. The prominence double layer is still more variable; indeed, it can explode, thus producing solar flares. A theory of solar flares, based on such current disruptions, has been worked out by Carlqvist (1969, 1973). He has presented a quantitatively consistent theory of solar flares.

Much effort has been spent on attempts to explain solar flares by a magnetic field formalism (a survey of these efforts is given by Heyvaerts (1979)). For reasons given in Chapter I (see Figure I.4), such an explanation is inadequate. To be more specific, since the boundary conditions are not correctly introduced in the magnetic merging theories of solar flares, these theories cannot explain the rapid concentration of the entire circuit's inductive energy at the point of disruption. On the other hand, there are theories which account for a solar flare as a disruption of a current, but attribute the disruption to instabilities other than exploding double layers. Such theories deserve to be taken seriously.

III.9. Solar Wind Acceleration

The solar wind is energized in the solar atmosphere, it is shot out into interplanetary space, and when it reaches the neighbourhood of the Earth it delivers part of its energy to the magnetosphere.

According to Chapter I (cf. Figure I.4) the transfer of energy from solar processes to the solar wind and from the solar wind to the magnetosphere can be described adequately only by a treatment of the electric current systems. Simple models for the circuits of energy transfer are given. Of these, Figure III.1(a) gives the basic mechanism for solar wind transfer of energy from the Sun to the Earth.

The plasma gun energizing and accelerating the solar wind at the Sun is represented by the accelerator region A, the motion through interplanetary space by the drift tube, and the transfer of some of the kinetic energy into the magnetosphere by the decleration region D. Of these three processes, the solar wind motion in interplanetary space has been studied by space probes and is reasonably well known. The transfer of energy into the magnetosphere has been analysed in some detail in III.6. In order to get a complete picture of the solar wind phenomenon, the solar processes remain to be studied. As these do not take place in a region which is now accessible to spacecraft we cannot hope to

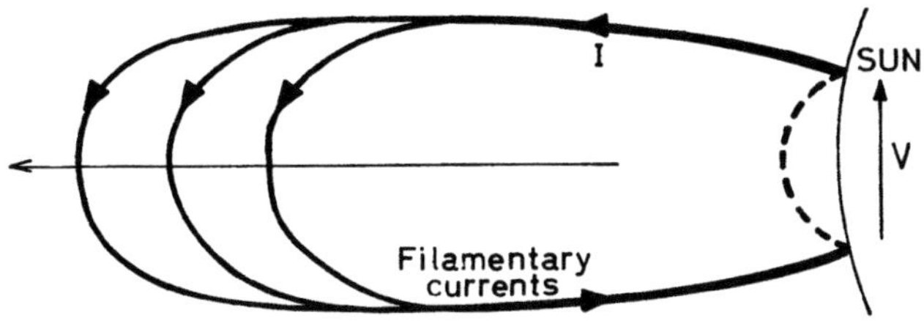

Fig. III.19. Currents are assumed to flow in quiescent prominences which occasionally erupt and given rise to expanding looplike structures in the corona. Due to the expansion the legs of the loops are transformed into coronal rays which carry currents from the photosphere to the outer parts of the corona or interplanetary medium and then back again to the photosphere.

understand them very well, but so much is known about the solar atmosphere that it should be possible to outline the basic processes.

Hence the problem to be discussed is how the solar plasma gun is constructed. We need an electromotive force represented by the battery in Figure III.1(a). This is of course given by the voltage difference V between two points A and B of the photosphere:

$$V = \int_A^B \mathbf{v} \times \mathbf{B}\, d\mathbf{s}. \tag{39}$$

Moreover we need conductors, transferring the power to the plasma gun, and these should presumably be magnetic field lines. Finally we need the plasma gun itself, a region where the circuit closes through currents perpendicular to the magnetic field, and hence accelerate the plasma.

The voltage difference may be due to motions in the neighbourhood of a sunspot, which according to III.8 is the voltage driving solar prominences. These are usually local phenomena, and cannot be expected to emit plasma into space. Exceptions are *rising prominences* which are seen to disappear into space at great heights. The voltage difference may also be produced by the solar rotation relative to a fix coordinate system (or the non-uniform rotation) and this should be a powerful source of energy in the polar caps, giving rise to currents along the field lines far out in interplanetary space.

The problem of energizing the solar wind by a system of electric currents has not yet been studied in a satisfactory way. An example of a rising prominence acting as a plasma gun has been briefly outlined by Carlqvist and Alfvén (1980). It is depicted in Figure III.19. The electromotive force in the photosphere produces a current which flows to great heights along one magnetic flux tube and down again along another. High up in the solar atmosphere (corona) the voltage difference transferred along the field lines accelerates the plasma in the solar atmosphere in the same way as in an ordinary plasma gun.

If the rich solar physics observational material is treated according to these principles, we may expect to be able to understand the basic mechanisms by which the solar wind is energized. When this is achieved we should be able to understand the whole mechanism of energy transfer 'from the solar photosphere to the terrestrial ionosphere'.

TRANSFER OF ENERGY FROM THE SOLAR CORE TO THE AURORA

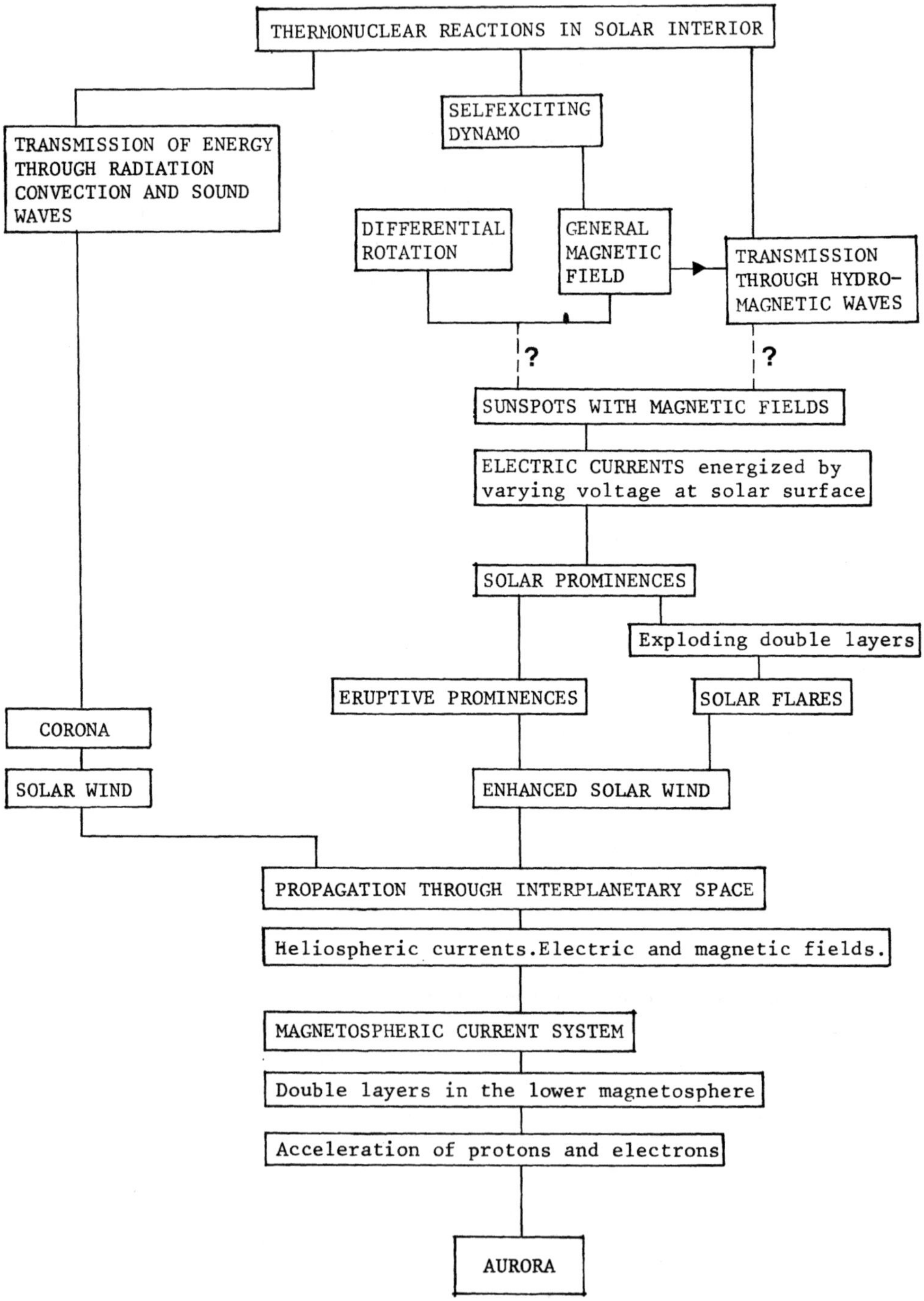

Fig. III.20.

III.10. Transfer of Energy from the Solar Core to the Aurora

Figure III.20 includes the transmission of energy from the solar core to the photosphere through various processes in the Sun. The processes are not well understood because *in situ* measurements are not possible, and also because confidence in the classical theory has been shaken by the neutrino difficulty. The transfer of energy from the photosphere to the solar wind is described in III.9, but again, unverified by *in situ* measurements. From that step in the transfer on, we are on somewhat firmer ground, and even if not all of the processes are very well studied, so much work is presently under way in exploring them that there is hope that we will understand the solar–terrestrial relations in the near future.

CHAPTER IV

THEORY OF COSMIC PLASMAS

IV.1. Classical Theory and Its Difficulties

The classical theoretical approach to cosmic plasmas can be characterized as the kinetic theory of gases generalized to include the electromagnetic forces acting on ions and electrons. The pioneers in the field were Chapman, Cowling and Spitzer, and their work has been further developed by a large number of prominent theoretical physicists. The result has been a mathematical formalism (which we shall refer to as the classical theory of cosmic plasmas) which is a very powerful tool in the hands of a competent scientist, but which, at the same time, is drastically misleading when not applied with sufficient care.

The reasons why it has often gone astray are essentially:

(1) Plasmas are extremely complicated. In order to derive the properties of a plasma theoretically (from first principles), it is necessary to make a number of simplifying assumptions. Particularly during the early development of the theory, these simplifications were not always appropriate. The lack of contact between theory and experiment allowed the theory to become excessively abstract, and in several respects, unconnected with reality. We have seen examples of this in the preceding chapters and will discuss some of these examples below.

(2) Real plasmas are very often 'noisy', i.e., they generate oscillations in a broad frequency band, leading to effects which are almost impossible to describe by the classical formalism. It is possible to generate noise-free plasmas in the laboratory (for example, externally heated cesium plasmas), but it seems that as soon as an electric current is sent through the plasma, it becomes noisy. Hence, what seems to be a basic property of a cosmic plasma has not yet been included in the classical theory. Consequently, there is little reason to expect that the classical theory could be generally valid. For example, the electron energy distribution in a plasma is very often non-Maxwellian, with the result that the high energy tail may ionize much more than expected.

(3) The application of classical theory to geophysical and astrophysical problems is often made in a careless and irresponsible way. Among the most common mistakes are:

(a) Infinite models are applied to plasmas with finite extension. For example, in the 'bow shock' (in the solar wind outside the magnetosphere) an electric current necessarily flows in the transverse direction. Of course, this current must close, but this point is neglected in most magnetospheric models with serious consequences (III.6.7).

(b) In theories about contracting interstellar clouds, it is often claimed that a magnetic field necessarily counteracts the contraction. This conclusion is model-dependent. The magnetic field can just as well assist, or even cause the contraction (see IV.8.3). Similar errors have led to the neglect of the pinch effect as an important mechanism for confining cosmic plasmas and producing filamentary structures (II.4).

(4) Very few cosmic scientists seem to be interested in a critical analysis of the

fundamentals of the theory of cosmic plasmas. Once a highly speculative model becomes 'generally accepted', it easily becomes sacrosanct.

The result of all this is that the theoretical treatment of cosmic plasmas is not in a very healthy state. It seems that a rather drastic and inevitably painful revision of the theory is necessary. Some of the reasons for this have been analyzed in Chapters II and III. In this chapter we shall discuss a number of other phenomena, which also motivate more or less drastic revisions of some aspects of the classical theory of cosmic plasmas.

IV.1.1. THE REVERSE DEFLECTION

As an example of the difficulties which the classical theory encounters even in an apparently simple configuration, the behaviour of a plasma flow in a longitudinal curved magnetic field will be discussed (Lindberg and Kristofersson, 1971; Lindberg, 1978).

As Lindberg states, on basis of simple theoretical arguments, the plasma may be expected to behave in a number of different ways when entering the curved region (see Figure IV.1). Depending on the parameters of the plasma (e.g., density, velocity, temperature and magnetization), it is anticipated that the plasma behaves as in Figure IV.1(b, c, or d). When Lindberg did the experiment, he found that for a certain parameter range, the beam deviated in the reverse direction (Figure IV.1(e)), and furthermore, the initially cylindrical beam contracted into a flat slab. Thus, *all* of the naive theoretical predictions were wrong! The same phenomenon has been observed by Tanaka *et al.* (1972) and Komori *et al.* (1977).

Lindberg demonstrated that *once observed*, the reverse deflection can be qualitatively understood on basis of classical theory. However, the phenomenon is rather complicated and the decisive factor responsible for the reverse deflection is the backward drift of high-energy electrons (see Figure IV.2). This phenomenon also illustrates the importance of the circuit aspect. The plasma first entering the region of curved field induces a transverse electric field $\mathbf{E} = -\mathbf{v} \times \mathbf{B}$. This area then acts as a generator driving current into the upstream region, which becomes polarized because of its low transverse conductivity. In conclusion, we can expect the reverse deflection to be a phenomenon of fundamental character which may occur in collisionless plasmas for a wide range of parameters.

As cosmic plasmas often move along curved field lines, the *moral* of the experiment is that without an intimate contact with experiments not even such apparently simple problems can be safely treated by the classical theory unless extensive work is done on investigating all relevant plasma phenomenon over different parameter ranges.

IV.2. Ionization

IV.2.1. IONIZATION BY LIGHT, BY PARTICLE RADIATION, AND BY ELECTRIC CURRENTS

Ultraviolet starlight is the main source of ionization in the Strömgren H II regions. In the H I regions it also ionizes elements with low ionization potentials. This has led to the belief that it is the *only ionizer* (in some cases supplemented by cosmic radiation) and that in dark clouds where starlight is absorbed (and also cosmic radiation, the ionization may cease, or in any case decrease to negligible values. This is not necessarily correct, because in many cases most of the ionization of a cosmic plasma is produced by a hydromagnetic conversion of kinetic or gravitational energy into ionizing electric currents.

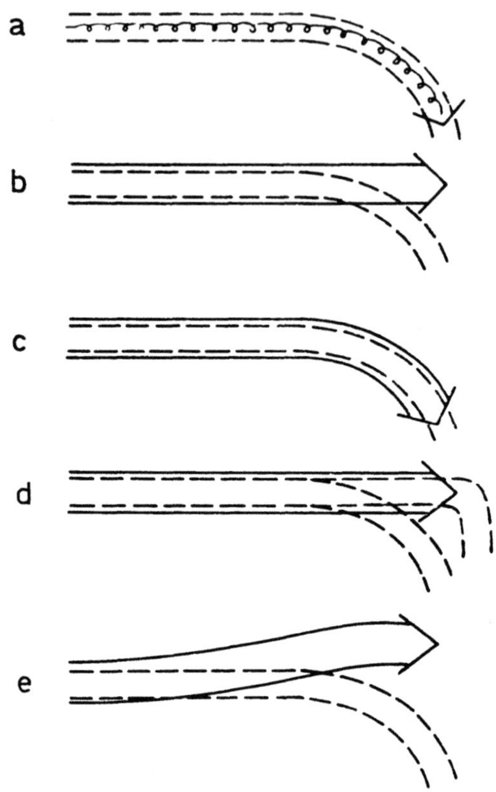

Fig. IV.1. Classical predictions and real behavior of a plasma beam, initially moving parallel to a magnetic field when entering a region of curved field (Lindberg and Kristofersson 1971). (a) Single particles follow in first approximation magnetic field lines even if they are curved. (b) A thin plasma beam becomes self-polarized and proceeds straight on by E × B-drift. (c) The polarization field is short-circuited by depolarization currents and the beam has to follow the curved field. (d) A plasma beam with high conductivity and kinetic energy moves straight on, 'stretching and pushing the magnetic field lines aside'. (e) Experiments with a collisionless plasma show a deflection contrary to all the theoretical predictions.

This is obvious from a study of our close surroundings. The solar light ionizes the dayside ionospheres of the planets, but with the exception of these very tiny regions, almost all ionization in the solar system is produced in other ways. The nightside ionosphere and the auroral zones are ionized partly by the impact of charged particles coming from the magnetosphere and partly from electric currents (e.g., the auroral electrojet). The magnetosphere is ionized by energy transferred from the solar wind, which in turn derives its kinetic energy, temperature, and ionization from the electric currents in the close environment of the Sun (III.8–10).

Such currents also heat the corona. The currents are energized by hydromagnetic processes which obtain their energy from convection in the Sun. This long chain of processes is almost completely independent of the light which the Sun emits. If the light emission (including ultraviolet light) of the photosphere could be switched off, the state of

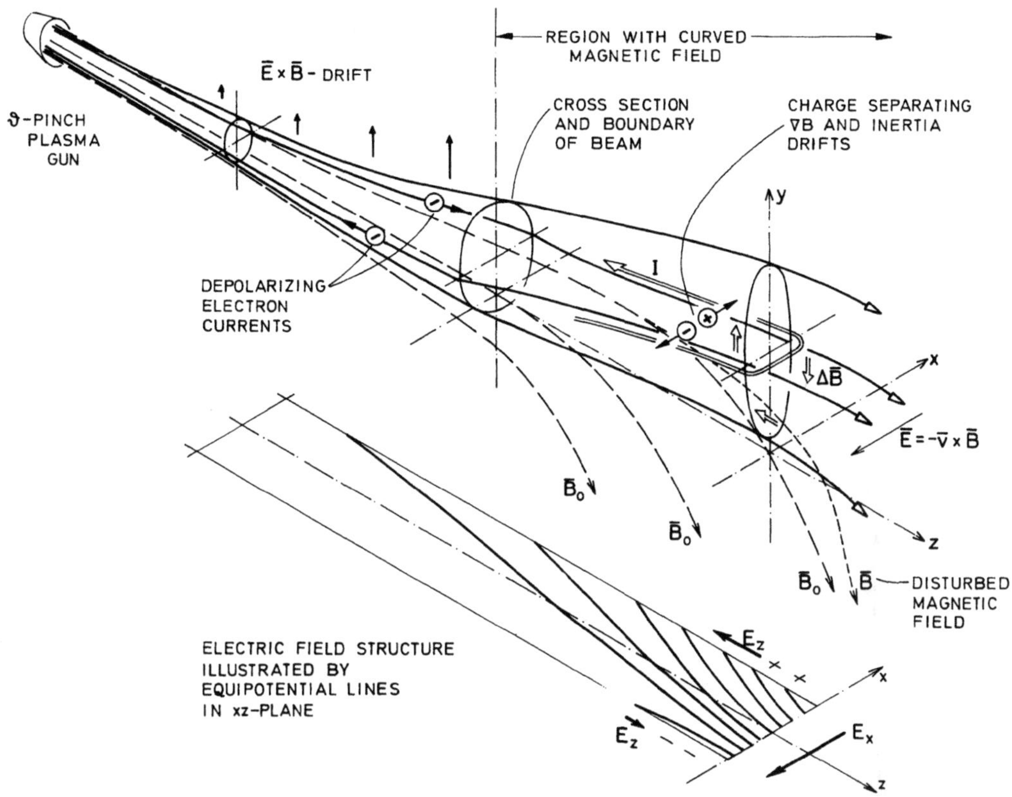

Fig. IV.2. Explanation of the unexpected behavior. Further investigations have clarified the cause of this. Plasma is shot from a theta-pinch plasma gun into a longitudinal magnetic field which downstream is curved downwards, B_0. When plasma enters the region with curved magnetic field charge separating particle drifts occur and a transverse electric polarization field $\mathbf{E} = -\mathbf{v} \times \mathbf{B}$ is set up. A fraction of the polarization field is propagated upstream due to backflow of electrons into the region with purely longitudinal magnetic field (depolarizing electron currents). This electric field gives an upwards $\mathbf{E} \times \mathbf{B}$-drift of the whole plasma beam before it enters the curved field. When entering the curved field it also contracts to a vertical flat slab. This is an effect of longitudinal electric field components. The potential distribution in the beam becomes unsymmetrical as shown below. This is an effect of the ions having much higher (translational) energy than the electrons. At high plasma densities the depolarizing currents change the magnetic field so that the field lines are first stretched (less tilted) and further downstream more tilted than in the undisturbed case (Lindberg, 1978).

ionization in the solar surroundings, including the upper parts of the solar atmosphere, would remain the same (with the aforementioned small exception).

Similarly, the high degree of ionization in the Jovian magnetosphere has very little to do with the light emission from the Sun or from Jupiter.

Hence, it is seen that very little of the ionization in our surroundings is produced by ultraviolet light (or cosmic radiation) but instead is generated by electric circuits which ultimately obtain their energy from the kinetic energy released in the solar convection zone (III.1.3 and III.9). Enough is known about this rather complicated chain of energy transfer to be able to follow it, at least in general, from the source of the energy to the ionization in our surroundings (III.10).

As an example of how the study of phenomena in our close neighborhood affects our views of more distant phenomena, we shall apply these conclusions to the formation of stars (IV.8). It should be noted that the formation of a star from a dark cloud represents a release of gravitational energy of the order of several thousand times the sum of the ionization energies of all the atoms in the whole mass.

As hydromagnetic transfer is, in general, a rather efficient process, it may be concluded that the processes discussed are able to produce a high degree of ionization during the formation. Although a quantitative analysis of the transfer processes is difficult to carry out in detail, it may be expected that there is enough ionization to exclude a non-hydromagnetic treatment of star formation. (This view is further supported by the fact that space to a large extent consists of different plasma regions, see II.9). Of course, this does not exclude the fact that there are partial processes, e.g., the drift of dust and the formation of stellesimals, which can be treated essentially without the introduction of hydromagnetic processes.

IV.2.2. TRANSITION BETWEEN A PLASMA AND A NEUTRAL GAS

When a region of highly ionized plasma is surrounded by a region of non-ionized gas, a transitional region of partially ionized gas is formed. The properties of such regions have been investigated by Lehnert (1968, 1970, 1975). The studies are based on results of thermonuclear research which are applicable to cosmic condition i.e., $n > 10^{13}$ m^{-3} and $B > 10^{-4}$ T. Such a range includes stellar atmospheres, including prominences and the lower solar corona.

Lehnert finds that the region of partially ionized plasma which separates the fully ionized plasma from a surrounding neutral gas, is often thin, so that there is a rather sharp border between the hot and the cool region. Figure IV.3 shows a typical case. The neutral gas is to the left, and the fully ionized region is to the right. In the boundary region, the neutral gas diffuses (short mean free path) into the boundary, and is essentially ionized by electrons from the plasma. The results may be applicable to the sharp transitions between active and passive plasma regions which are observed in many cosmic situations.

IV.3. Cosmic Abundances and Differentiation

IV.3.1. COSMIC ABUNDANCES

For a long time there has been a general belief that the 'cosmic abundances' of the different elements were almost the same everywhere in the Universe. There were, of course, exceptions like some planets and satellites, for example the Earth and the Moon, whose gravitation was insufficient to keep the most volatile substances, and in which geological processes had produced segregations. The chemical and petrographic differences between meteorites were attributed to similar phenomena in the 'broken-up planet' (or planets) in which segregation had taken place and from which the meteorites were believed to originate. Hence, not very long ago it was claimed that Jupiter (and sometimes the other giant planets also) must necessarily consist of 'solar matter' or even of solid hydrogen, a view which still seems to be held by many of the scientists who work on the origin of the solar

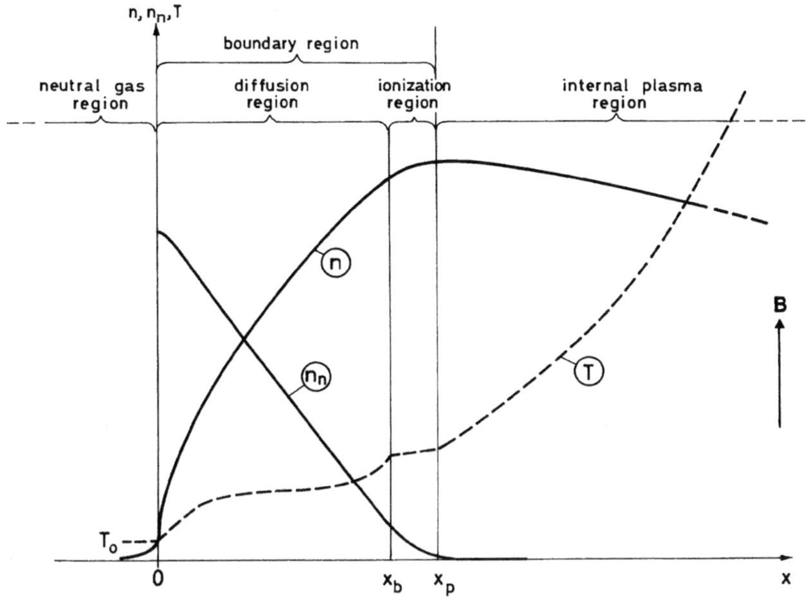

Fig. IV.3. Transition between a fully ionized plasma and a neutral gas. The neutral gas region is defined by $x < 0$, the boundary layer by $0 < x \leqslant x_p$, and the fully ionized region by $x > x_p$ (Lehnert, 1975).

system. The motivation for such beliefs was that turbulence had mixed the matter in the Universe, so that no variations in the abundance ratios were possible. The fact that some stars showed chemical 'anomalies' was attributed to 'anomalies' in the nucleo-synthesis in their interiors. In any case, in the plasma of the solar system existing now or in the early history of the solar system, there could be no chemical differences because the plasmas were in a turbulent state. When it was pointed out that the band structure of the solar system (V.7) required clouds of different composition falling in towards the Sun, this was considered to be fatal to the concept of a band structure.

All this changed drastically when *in situ* measurements in space became possible. Like so many other homogeneous models, the belief in chemical homogeneity did not endure an encounter with observational facts.

IV.3.2. OBSERVATIONS OF CHEMICAL DIFFERENTIATION

The first serious objection to chemical homogeneity came when solar wind measurements showed that its helium content was variable. From an average value of a few percent it sometimes, especially in connection with solar flares, went up to more than 20% and sometimes down to lower values (Hirschberg, 1973). Attempts to explain this as due to nucleo-synthesis in the flares did not look very attractive. However, the decisive demonstration of very large chemical differentiations came from investigations by the Johns Hopkins group into the chemical composition of the low energy solar cosmic rays emitted in connection with solar flares. A review of their results is given by Zwickl, Roelof, Gold, Krimigis, and Armstrong (1978). See also Briggs, Armstrong, and Krimigis

(1979). It is especially important to note that their results indicate that the chemical differentiation was not produced *by the flare* itself, but by processes working *before the flare*. The argument for this is essentially that the flare-produced energy spectra of the different components are the same. Hence their results indicate that *there are processes in the solar plasma which produce a chemical separation under pre-flare conditions*.

IV.3.3. SEPARATION DUE TO DIFFERENT IONIZATION POTENTIALS

As a solar flare is due to a current disruption, there are indications that it is an electric current which causes the differentiation, possibly in combination with gravitation. If a filamentary current flows in a partially ionized plasma it will cause all the ions to move towards the axis.

We have discussed this in II.4.5 and have concluded that the formation of magnetic ropes is a possible mechanism for producing regions with different chemical composition.

There is also another somewhat related general process which is known to produce chemical differentiation. During his thermonuclear experiments with a fully ionized magnetized plasma surrounded by a 'blanket' of non-ionized gas, Lehnert (1968, 1976) investigated the boundary region between the fully ionized and the non-ionized gas. He found that if the plasma consists of a mixture of different elements a chemical differentiation will take place. This causes an enrichment of the elements in one of the regions (in an ionized or non-ionized region, depending on whether their ionization potentials are high or low). Under certain conditions the enrichment factor may be so large that the anomalies in the solar wind (and also in the magnetic stars) may be explained.

IV.3.4. SEPARATION DUE TO MASS DIFFERENCES

In III.1.2 we have found that some of the guiding center drift terms are mass dependent. One of them is the gravitation drift produced by $\mathbf{F}_g = m\mathbf{g}$. If the plasma is 'hung up' in a magnetic field or is supported by a gas pressure, the gravitational drift can be exactly compensated, but only if all of the ions have the same mass. If the plasma contains a mixture of particles with different masses, non-mass dependent forces can only compensate an average value of F_g. If this occurs, the light ions will have a net motion in one direction and the heavy ions will have a motion in the other. It is not certain, however, that this effect is quantitatively large enough to be of importance.

Another process which produces a separation of particles having different masses has been suggested by Fisk (1978). There are probably quite a few other processes which could produce mass separation.

IV.3.5. SEPARATION DUE TO DIFFERENT VAPOR PRESSURES

When considering plasma processes for chemical separation we should, of course, not forget that in dusty plasmas, the plasma condenses into grains. The evaporation of these grains is often an important differentiation process.

In addition, radiation pressure may in some cases contribute to chemical separation.

IV.4. Turbulence

Turbulence in a fluid is generally defined as a state where the velocity of the fluid is truly *random and disordered* (see, e.g., Tsytovich, 1977). This means that the instantaneous velocity measured at any point in the turbulent region is non-reproducible. It is often claimed that turbulence is a very important phenomenon in cosmic plasmas. The study of the magnetosphere by *in situ* measurements has not given any indication that *large scale turbulence* is of importance. On the contrary, modern models exclude the existence of large scale turbulence (Heikkila, 1972).

Certainly spacecraft measurements of density, magnetic field, and plasma velocity often show rapid variations. This is especially true when a spacecraft passes magnetic field lines going through the auroral zone, the magnetopause, or the neutral sheet in the tail. However, these variations usually seem to be due to current filaments or current sheet often forming a regular, obviously non-turbulent pattern.

The region of the terrestrial atmosphere up to the turbopause at about 100 km is often in a highly turbulent state. At the still greater heights of the ionosphere, we often find – especially in the auroral zone – rapid and irregular variations. In part, these are due to convection produced by local heating, but at least in the upper ionosphere, they seem mostly to be due to rapid changes in the current pattern, and the secondary effects of this. It is misleading to identify them as due to 'turbulence' in the proper sense of the word.

The *solar atmosphere* is generally characterized by filaments, e.g., prominences, spicules, and coronal streamers. If a spacecraft could be sent through these regions, it would register rapid and seemingly irregular variations in density, temperature, and magnetic field. From a study of the records one would be tempted to conclude that the upper solar atmosphere were in a turbulent state. Visual observation of the filamentary structure contradicts this interpretation. Certainly the filamentary structure often changes rapidly, but not in a way which makes it possible to speak about 'turbulence' in the proper sense of the word.

However, it is quite possible that the convective layer in and below the photosphere can be described in terms of turbulence, although the convection is often structured in a way which shows that it is not really random.

The so-called 'shock front' on the sunward side of the magnetosphere is often thought to be in a turbulent state. There is no doubt that spacecraft traversing it register irregular and rapid variations in density, temperature, and magnetic field. When passing the shock front, the solar wind slows down, and as it consists of a collisionless plasma, this can be done only by electromagnetic forces acting on the particles. There is no doubt that in the shock front region there is a network of perhaps rapidly varying electric currents which interact with the magnetic fields partially produced by themselves, but it is not at all evident that this should be referred to as 'turbulence' (III.6). In fact, it is misleading to use this term for a phenomenon which has very little similarity with, for example, tropospheric turbulence.

The distinction between 'turbulence' and 'rapid apparently random fluctuations' is not only a semantic question. Turbulence is usually associated with *mixing*. In the turbulent region below the turbopause there cannot be any appreciable difference in chemical composition because of the turbulence-induced mixing (see, e.g., Rishbeth and Garriott, 1969). Similar mixing does not necessarily take place in a region with a filamentary structure. On the contrary, there is evidence for the opposite.

As we have seen in IV.3.2 the He/H ratio in the solar wind and the chemical abundances in the emitted low energy solar cosmic rays may vary drastically (between

$< 1\%$ and $> 20\%$ for the He/H case). This is likely to be due to a chemical separation by plasma processes (e.g., due to differences in ionization potential). Such processes are, of course, impossible if the regions from where the wind originates were in a turbulent state with associated mixing.

There are many phenomena in cosmic physics which are attributed to turbulence. For instance the line broadening observed in different celestial objects is often interpreted in such a way. As we have found, there is no clear evidence that turbulence is of importance in the low density space plasmas which so far have been explored by *in situ* measurements. On the contrary, it seems that magnetic fields give a structure to low density plasmas which suppresses or prevents turbulence. A perturbation is more likely to produce hydromagnetic waves than turbulence. In fact it seems that most of the fluctuations in the magnetosphere and solar wind can be accounted for as a superposition of hydromagnetic waves (Belcher and Davis, 1971; Hollweg, 1975).

It is reasonable to conclude that there is also no pronounced large scale turbulence in those regions not yet accessible to *in situ* measurements.

There are theories attributing the production of cosmic magnetic fields to turbulence. There is also a widespread belief that there is so much turbulent mixing in the Universe that the 'cosmic abundances' of the elements must be the same everywhere. *Speculations of this kind* must *not be taken seriously unless decisive arguments for the existence of large scale turbulence are presented*. The careless habit of using 'turbulence' for 'apparently random fluctuations' is often misleading and should be abandoned.

However, even if we see no evidence for large scale turbulence, there is decisive evidence that what is often called *small scale plasma turbulence* exists. In a survey of 'Plasma Turbulence', Kodomtsev (1965) points out that when applied to a laboratory plasma, the term 'turbulence' is used in a broader sense than in conventional hydrodynamics. The 'turbulence' which normally occurs in a plasma causes noise and oscillations, and is responsible for the 'anomalous' Bohm diffusion. A weak turbulence of this kind shows greater similarity to the motion of the wavy surface of the sea, than to the turbulent motion of an ordinary fluid. In other words, it does not necessarily produce very much chemical mixing.

Finally, even if there is no turbulence in the proper sense of the word, interfering plasma waves, especially if they have large amplitudes, may produce considerable mixing in certain regions. In conclusion, we badly need a partially semantic analysis of the role of 'turbulence' in cosmic plasmas.

IV.5. Flux Amplification

IV.5.1. PRODUCTION OF COSMIC MAGNETIC FIELDS

As we have seen in the first three chapters, an understanding of cosmic hydromagnetic phenomena is impossible without describing them *explicitly* by electric currents. This also applies to the production of cosmic magnetic fields.

There are at least three different types of current systems which produce cosmic magnetic fields:

A. *Current Loops in the Interior of Celestial Bodies*

The magnetization of celestial bodies must be due to electric currents flowing in their interiors. This will be treated in this section.

B. *Currents Near the Surface*

Examples: Ionospheric currents in the upper atmosphere of the Earth, currents associated with Sunspots and prominences in the Sun.

C. *Large-scale Current Systems*

Examples: Magnetospheric current system, heliospheric current system, equatorial plane current around Jupiter, and currents in the cometary tails, probably also a general galactic current system.

IV.5.2. SELF-EXCITING DYNAMOS

It seems possible to account for the current systems B and C mentioned above, as generated by the e.m.f.'s V produced by hydrodynamic motions v in the presence of a magnetic field B. We have

$$V = \int \mathbf{v} \times \mathbf{B}\, \mathrm{d}s. \tag{1}$$

However, a primary field B must be postulated, and this must derive from another mechanism. The integrated process is traditionally referred to as 'self-exciting dynamo'.

Self-exciting dynamos must exist in the interior of some celestial bodies like the Earth, Jupiter, Saturn and the Sun. Also, it is impossible to avoid galactic dynamos, most probably located in the nuclei of the galaxies.

Dark interstellar clouds are known to be magnetized, with fields as large as 10^{-7} T (10^{-3} G) or even more. A self-exciting dynamo is likely to exist in the galactic core. The field in dark interstellar clouds may be concentrations of this galactic field. However, the stronger field in dark clouds may just as well be generated by self-exciting dynamos of their own.

Furthermore, there is a very interesting, although little studied 'transplanetary region' which at the time of the formation of the solar system was important because it was part of the 'source cloud' (ESS, Ch. 19). Also under present conditions there may be a similar region which is intermediate between the heliosphere (which is dominated by the solar wind) and what may be called true interstellar space. This is the region which may be related to the Oort cometary reservoir and may serve as a reservoir of what are usually called 'galactic' cosmic rays (which is likely to be a misnomer, because there is no convincing argument for the view that they all are galactic (IV.10.3)). If this region should serve as a cosmic ray reservoir we may need a self-exciting dynamo even here, in order to produce enough magnetization.

So our conclusion is that self-exciting dynamos are likely to exist not only in the interior of some celestial bodies, but also in some space plasmas.

IV.5.3. THEORIES OF SELF-EXCITING DYNAMOS

There are a large number of well-known theories of self-exciting dynamos. Space dynamos have been attributed to plasma turbulence, but as we have seen in IV.4, there is no convincing evidence that turbulence in the real sense ever exists in cosmic plasmas. Even the usual models of dynamo processes in the interiors of celestial bodies seem to encounter difficulties (see Srnka and Merrill, 1979) so that alternatives to those should also be considered.

IV.5.4. A FLUX AMPLIFICATION MECHANISM

We shall now discuss a somewhat different approach which does not rely on more or less speculative processes. It starts from a process which is known from laboratory experiments. to lead to an amplification of an already existing magnetic field. The experiment in which this mechanism was discovered was carried out by Lindberg *et al.* (1960), and Lindberg and Jacobsen (1961).

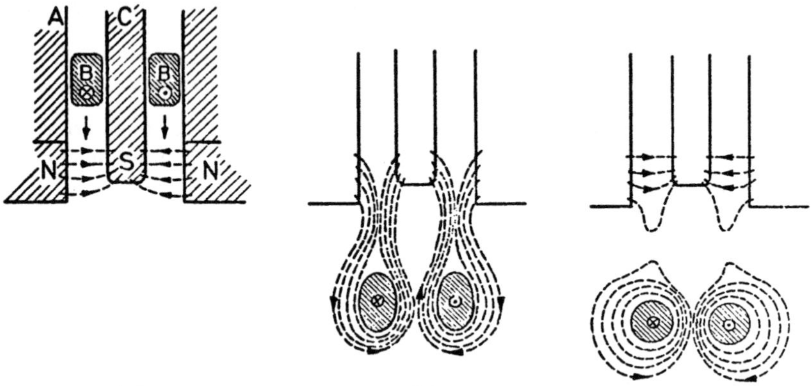

(a) (b) (c)

Fig. IV.4. Production of a plasma ring with both poloidal and toroidal magnetisation. (a) A coaxial plasma gun produces a plasma ring with toroidal magnetisation. When leaving the gun it passes through a radial magnetic field. (b) The ring 'captures' the field so that it gets a poloidal magnetisation. (c) Free ring with both poloidal and toroidal magnetisation (Lindberg *et al.*, 1960).

A plasma gun produces a ring of plasma (Figure IV.4) which because of the discharge current in the gun has a toroidal magnetization. At the muzzle of the gun there is a radial magnetic field. This radial field is captured by the plasma ring so that the plasma ring also acquires a poloidal magnetization. The behavior of the ring is studied when it has drifted away from the gun.

Hence we have a free plasma ring with both toroidal and poloidal magnetization. It is found that if the toroidal magnetic energy is larger than the poloidal energy there may be a transfer of magnetic energy from the toroidal to the poloidal magnetic field so that the poloidal magnetic flux is amplified.

From detailed diagnostics of the plasma, it is shown that the flux amplification can be easily understood *if we depict the current system* (Figure IV.5). In simplified representation the poloidal field B_p is produced by a ring current I_p. The toroidal magnetic field B_t is produced by an axial current I_t which closes at large distances. Under certain conditions the axial current exhibits a kink instability, causing the current to form a spiral. The new current configuration can be described as the sum of an axial current and a circular loop. Due to the properties of the kink instability, the circular loop is always oriented so that it amplifies the already existing poloidal field.

The phenomenon of flux amplification which has been studied in the laboratory *may explain the production of a magnetic field in low density plasmas in cosmic physics* (Alfvén, 1950, 1961; CE; Alfvén and Lindberg, 1974).

Plasma phenomena in similar geometries have been studied in reverse field pinches and speromaks (Fowler and Coensgen, 1979; Goldenbaum *et al.*, 1979).

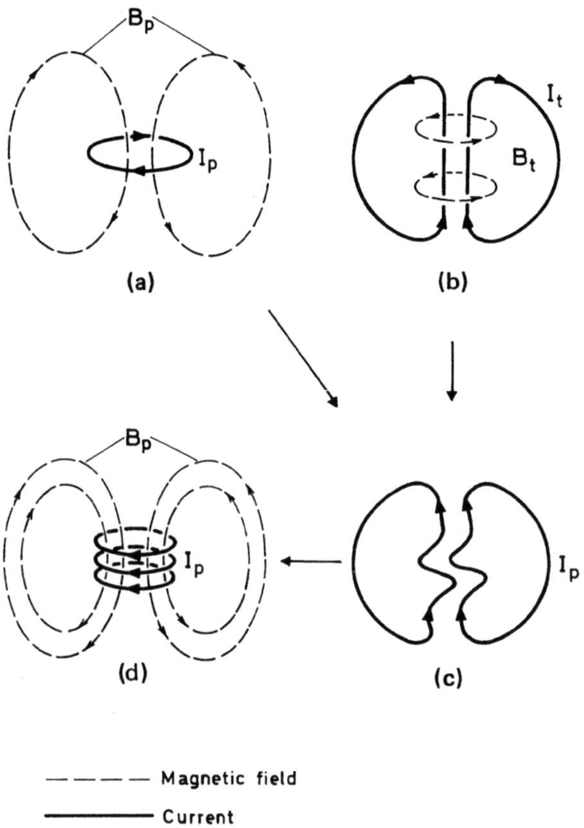

Fig. IV.5. Flux amplification in Lindberg's experiment (Lindberg and Jacobsen, 1961). Original magnetization of the plasma ring has a poloidal and a toroidal component. (a) Poloidal field produced by ring current. (b) Toroidal field produced by axial current which is unstable. (c) Kink instability of axial current produces ring currents. (d) Resulting amplified poloidal field.

IV.5.5. PRODUCTION OF COSMIC MAGNETIC FIELDS BY THE KINK INSTABILITY

Whether kink instability flux amplification also can explain the origin of magnetic fields in interstellar clouds is not certain. Regarding planetary interiors it is even more of an open question. But should this be the case, it may be done in the following way:

(1) There exists a primary poloidal field (however weak) (Figure IV.6(a)).

(2) Through hydrodynamic motion the 'frozen in' magnetic field lines are deformed in such a way as to produce a toroidal field. For example, suppose that the rotational velocity in the central region of the equatorial plane differs from that in the outer regions. Two toroidal magnetic fields with opposite directions in the northern and southern hemispheres are produced (Figure IV.6(b)).

(3) When the energy of the toroidal magnetic field is comparable to the poloidal energy, the axial currents associated with this become unstable (Figure IV.6(c)). The kink instability causes ring current components to be produced; and these ring currents always amplify the original poloidal field (Figure IV.6(d)).

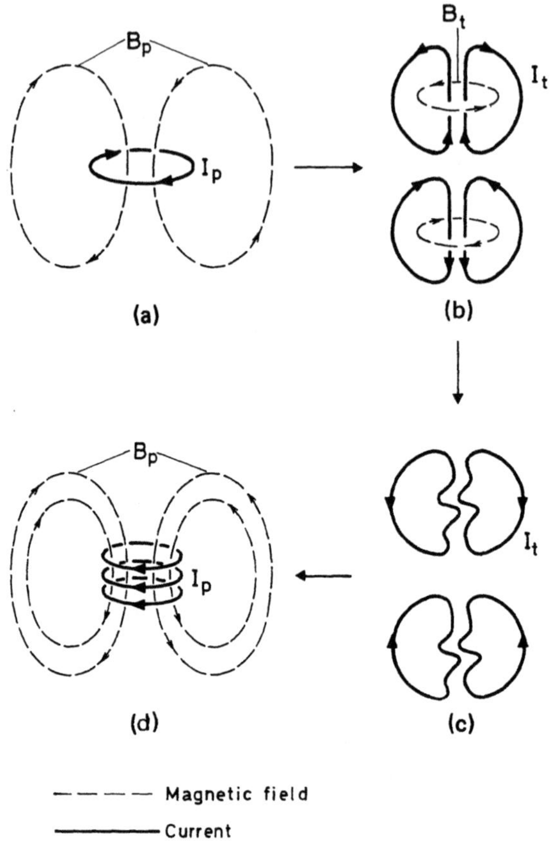

Fig. IV.6. Kink instability theory of cosmic flux generation. (a) Original poloidal magnetic field produced by ring current. (b) Slowdown of equatorial rotation deforms magnetic field lines so that two toroidal field components are produced. (c) Both toroidal fields are associated with axial currents which produce kink instabilities. (d) Resulting amplified poloidal field (Lindberg and Jacobsen, 1961).

(4) Iterations of this process may lead to very strong magnetic fields. A limit is given by the kinetic energy available in the differential motion.

The theory is basically non-speculative. Conditions (1) and (2) are known to generally exist under cosmic conditions. Condition (3) is an application of the best-known of all hydromagnetic instabilities. Through (2), kinetic energy is converted into toroidal magnetic energy. Through (3) this is converted into poloidal energy.

It is an open question whether the conditions in planetary interiors are such as to make the theory applicable. In view of the obvious difficulties of all theories of magnetic flux generation in planets a serious effort should be made to clarify this.

IV.6. Critical Velocity

IV.6.1. PREDICTION OF THE CRITICAL VELOCITY IN THE EARLY SOLAR SYSTEM

If we plot the gravitational energy W_g of the bodies in the solar system (planets and satellites) against their distance to the central body (sun or planet) around which they move, we find that all of the 'regular' bodies (i.e., excepting retrograde satellites and asteroids) fall in certain bands. This very pronounced band structure of the solar system (V.7) can be explained if non-ionized matter falling in from large distances becomes ionized and hence stopped by electromagnetic forces as soon as it has reached the 'critical velocity' v_c defined by

$$\frac{mv_c^2}{2} = eV_{\text{ion}} \tag{2}$$

(m = atomic mass and eV_{ion} = ionization energy of the gas).

When the band structure of the solar system was found (Alfvén, 1942), it was first thought that it would be very easy to explain: as soon as the atom had accumulated the energy needed for ionization, it became ionized. However, attempts to derive this from classical plasma theory showed that it was *not* expected theoretically for a number of reasons: it was not easy to find a process which transferred the kinetic energy of the atoms (not the electrons!) into ionization. Further, of the available energy only a rather small fraction could be transferred into ionization because of a number of competing processes. Indeed it was impossible to explain the band structure theoretically.

IV.6.2. EXPERIMENTAL DISCOVERY

However, as classical plasma theory should not be considered sacrosanct, it was essential to investigate the phenomenon experimentally. As soon as thermonuclear research made experimental techniques available, an investigation was made into how a magnetized plasma interacted with a non-ionized gas. These experiments led to the *experimental discovery of critical velocity*. In other words, the conclusions drawn from the band structure of the solar system were correct, and the classical theory was not!

After the initial experimental discovery of the critical velocity, the phenomenon has been demonstrated in a number of different configurations: homopolar rotation experiments, coaxial plasma guns, and in Lehnert's thermonuclear experiment with a rotating plasma. The experiment which is not directly applicable to the cosmic problem is Danielsson's experiment where a magnetized plasma is shot into a cloud of non-ionized gas (Danielsson, 1970).

It has also been shown that a 'voltage limitation', found experimentally long before the experiments mentioned above were done, could be explained by the critical velocity phenomenon. A survey of the experimental investigations has been given by Danielsson (1973).

More recent experiments have been done with more sophisticated techniques by the Bochum group (cf. Himmel *et al.*, 1976, 1977; Möbius *et al.*, 1979a). In general they confirm the earlier results and open new possibilities for a deeper understanding.

All of these experiments were done with pure gases, but experiments by Axnäs (1978a, 1978b) have clarified how gas mixtures and molecular gases react in a number of cases.

The most remarkable fact uncovered by the experiments is that Equation (2) is valid over a large range of parameters. In fact, magnetization and degree of ionization can be varied, often over several orders of magnitude and still Equation (2) is in a reasonably good agreement with observation.

IV.6.3. THEORY

The theoretical explanation of the results of experiments on the critical velocity has turned out to be very difficult. A number of mechanisms have been suggested in order to explain the experimental results, but none of these seems to be able to cover more than a certain range of parameters. In a survey of the theories, Sherman (1973) claims that this has to be accepted. The transfer of energy may be caused by different mechanisms in different parameter ranges. The basic fact seems to be that a plasma is able to transfer kinetic energy from the ions to the electrons in a number of efficient ways.

Recent theoretical work has been done by Raadu (1975, 1978) who develops Sherman's approach and more specifically, focuses attention on the role of electrostatic instabilities. Varma (1978) and Piel *et al.* (1980) have considered mechanisms for electron heating depending on the formation of electrostatic sheaths of the type originally proposed by Lehnert (1967).

IV.6.4. THE CRITICAL VELOCITY AND SPACE RESEARCH

In spite of the mentioned difficulties, a fairly good theoretical understanding of the critical velocity phenomenon has now been achieved. This means that the time is ripe for explaining a number of astrophysical phenomena by this mechanism. Besides the band structure of the solar system there are quite a few phenomena to which it is applicable. As soon as the relative velocity between a magnetized plasma and a non-ionized gas exceeds v_c, the interaction between them should be strongly enhanced.

For the ionospheric gases N and O, v_c is about 11 km s^{-1}. This is just above the velocity of a satellite striking the upper ionosphere, so that an experiment from a satellite would require that a gas be ejected in the forward direction. An interesting spacelab experiment, involving the release of a neutral gas (xenon) with critical velocity less than the orbital velocity, to demonstrate the phenomenon has been suggested by Möbius *et al.* (1979b). Freeman *et al.* (1972), Lindeman *et al.* (1974), Gold and Soter (1976), and Srnka (1977) have all invoked critical velocity to explain various phenomena occurring at the lunar surface. There should also be implications involving the interaction between the solar wind and the ionosphere of Venus or a comet (Mendis and Ip, 1977). Furthermore, it seems reasonable that critical velocity plays an important part in the interaction between the solar wind and the interstellar medium (Petelski *et al.*, 1980), as well as in quite a few more distant astrophysical phenomena.

A recent review, covering both experiments, theory and space applications, is given by Raadu (1980).

IV.7. Dusty Plasma

Cosmic plasmas are very often 'dusty', by which we mean that they contain solid particles, some of which are very small dust grains. These are usually electrically charged, and if their charge to mass ratio is large enough their motion may be essentially controlled by electromagnetic forces so that they can be considered as part of a 'dusty plasma'. For a review, see Mendis (1979). Charged dust in planetary magnetospheres has been investigated by Hill and Mendis (1979, 1980).

The temperature of the solid particles in a cosmic plasma may differ from the plasma temperatures by orders of magnitude. If the plasma is transparent and the solid particles radiate into space, their temperature may, for example, be 10 K even if the electron temperature is 10^4 K, the ion temperature 10^3 K and the molecular temperature 100 K.

IV.7.1. SOLID PARTICLES AS PART OF A PLASMA

A solid particle in a cosmic plasma receives negative electric charge essentially from the plasma electrons hitting it. The particle may then lose the charge by the photoeffect, sometimes also by field emission, and by positive ions hitting it. Usually the particle charge is 1–10 V positive or negative. However, if the plasma contains superthermal electrons these can sometimes give the grain a negative potential of several thousand volts (DeForest, 1972; Reasoner *et al.*, 1976; Mendis, 1979). Sudden changes between a few volts and several thousand volts have often been measured on spacecraft orbiting the Earth.

IV.7.2. ELECTROMAGNETICALLY AND GRAVITATIONALLY CONTROLLED MOTION OF SOLID PARTICLES

If a body has an electric charge q, it is subject to an electromagnetic force

$$\mathbf{F}_E = q(\mathbf{E} + \mathbf{v} \times \mathbf{B}) \tag{3}$$

where **E** and **B** are the electric and the magnetic fields, and **v** the velocity.

If F_E is large compared to other forces the grain will spiral in the cosmic magnetic field. If the Larmor radius of this motion is small compared to the extension of the plasma, the particle can be considered as part of the plasma. Such *dusty plasmas* are very common in space. Dark interstellar clouds consist of dusty plasmas, and there are strong arguments for the view that the solar system was once formed out of a dusty plasma.

If a dusty plasma contains solid particles with an extended mass spectrum, there exists a limit, separating those grains which are small enough to be part of the plasma, and those which are big enough not to be strongly influenced by electromagnetic effects. The size limits depends on the magnetic field, the electric field, the gravitation, the charge and the mass density of a particle. The limit is often around 10^{-5}–10^{-7} m. (For a survey, see ESS,5,4.) When a particle enters a region of plasma with a large quantity of superthermal electrons it may suddenly increase its charge by a factor of perhaps 1000, which may result in the capture of the particle in the plasma. If later its charge goes back to a low value, and if it accretes enough mass in the plasma, its motion may again be determined essentially by non-electromagnetic forces.

IV.8. Formation and Evolution of Interstellar Clouds

It is often believed that hydromagnetic effects counteract the contraction of a cosmic cloud. After what has been said in II.4, the opposite can just as well be true. It is easy to find magnetic configurations in which the contraction is assisted by electromagnetic forces. Such forces may even be decisive for the formation of cosmic clouds, and for keeping them together.

As we have found in II.4.3, the pressure gradient ∇p which is produced by stationary electromagnetic phenomena is

$$\nabla p = \mathbf{i} \times \mathbf{B}. \tag{4}$$

As

$$\mathbf{i} = \frac{1}{\mu_0} \nabla \times \mathbf{B} \tag{5}$$

we can eliminate either **i** or **B**. In cosmic plasma physics **i** is traditionally eliminated which gives

$$\nabla p = \frac{1}{\mu_0} (\nabla \times \mathbf{B}) \times \mathbf{B} \tag{6}$$

or

$$\nabla (p + B^2/2\mu_0) = \frac{(\mathbf{B}\nabla)\mathbf{B}}{\mu_0}. \tag{7}$$

Under certain assumptions (for example, if all magnetic field lines are parallel) the right hand term is zero and we have

$$p + B^2/2\mu_0 = \text{const.} \tag{8}$$

The general belief that a magnetic field counteracts the contraction of a cloud is essentially based on this formula, which is valid in only one special case. In order to illustrate the importance of the right-hand term we shall treat a simple case in more detail (Rose and Clark, 1961).

Consider a cylindrical coordinate system $(r, \phi. z)$ and suppose that the variables (pressure p, magnetic field **B**, and current density **i**) depend on r. Then we have from Equation (4).

$$\frac{\mathrm{d}p}{\mathrm{d}r} = i_\phi B_z - i_z B_\phi. \tag{9}$$

With $B_\phi = \mu_0 I_z(r)/(2\pi r)$, where $I_z = I_z(r)$ is the total current in the z direction inside a circle with the radius r, and from Equation (5)

$$i_\phi = -\frac{1}{\mu_0}\frac{\mathrm{d}B_z}{\mathrm{d}r}, \qquad i_z = \frac{1}{2\pi r}\frac{\mathrm{d}I_z}{\mathrm{d}r} \tag{10}$$

we find from Equation (7)

$$\pi r^2 \frac{\mathrm{d}}{\mathrm{d}r}\left(p + \frac{B_z^2}{2\mu_0}\right) = \frac{-\mu_0}{8\pi}\frac{\mathrm{d}(I_z^2)}{\mathrm{d}r}. \tag{11}$$

After integration by parts we obtain

$$-\pi r_1^2 \left(p + \frac{B_z^2}{2\mu_0}\right)_{r=r_1} + \int_0^{r_1} \left(p + \frac{B_z^2}{2\mu_0}\right) 2\pi r\,\mathrm{d}r = \frac{\mu_0}{8\pi} I_z^2\,(r_1). \tag{12}$$

IV.8.1. THREE SPECIAL CASES

We have three important special cases:

(1) $i_z = 0$. The current flows only in the ϕ-direction (see Figure IV.7(a)) and Equation (11) is satisfied by

$$p + \frac{B^2}{2\mu_0} = \text{const.} \tag{13}$$

The sum of the gas pressure and the 'magnetostatic pressure' is constant in space.

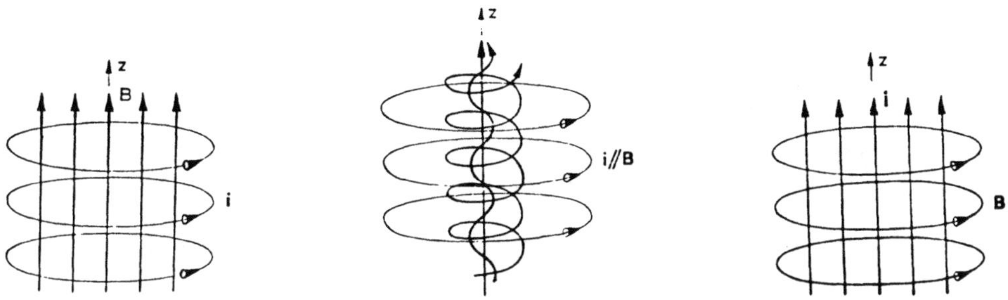

Fig. IV.7. Three special cases of stationary and cylindrically symmetric current (i) and magnetic field (B) configurations. (a) A toroidal current and an axial magnetic field leading to a force opposing contraction. (b) A force-free configuration with i and B parallel. (c) The Bennett pinch with an axial current and a toroidal magnetic field.

(2) $p = \text{const}$. The field is force-free (II.4.4). From Equation (9) we obtain

$$\frac{i_z}{i_\phi} = \frac{B_z}{B_\phi}. \tag{14}$$

This leads to a spiralized magnetic field (Figure IV.7(b)) which is parallel to the z-axis at $r = 0$, but perpendicular to it at large r-values. For large r-values B_ϕ decreases in proportion to r^{-1} (Murty, 1961).

(3) The third case is given by the conditions $B_z = 0$ and $p(r_1) = 0$. This is equivalent to the classical pinch with an azimuthal magnetic field and currents in the z-direction (Figure IV.7(c)). In a fully ionized plasma having the temperature $T_i = T_e = T$ and the density $n_i = n_e = n$ and with $N = \int_0^{r_1} (n_i + n_e) 2\pi r \, dr$ (= total number of ions and electrons per unit length) we obtain from Equation (12) the Bennett relation (Bennett, 1934)

$$\frac{\mu_0}{4\pi} I_z^2 = 2NkT \tag{15}$$

earlier mentioned in II.4.3.

Even if the gas pressure is zero outside r_1 we can confine a plasma inside r_1.

When a plasma is only partially ionized, the electromagnetic forces act on the non-ionized components only indirectly through the viscosity between the ionized and non-ionized constituents. If the coupling is strong we should add a term $2N_A kT_A$ in Equation (15) (the subscript A refers to non-ionized components).

IV.8.2. FORCE-FREE MAGNETIC FIELDS AND THE PRODUCTION OF FILAMENTS

This case has been treated in II.4.4. Let us now consider a plasma which is penetrated by a nearly force-free magnetic field and current forming a spiral structure similar to that shown in Figure IV.7(b). The plasma is assumed to have a high but not infinite conductivity. This implies that an electric field E_z has to be present. At a large radial distance r from the axis the magnetic field is almost toroidal and of the magnitude $B_\phi = \mu_0 I_z/2\pi r$. I_z is as before the total current in the axial direction inside r. Under the influence of the electric field there is an inward radial drift

$$|v_r| = \frac{E_z}{B_\phi}. \tag{16}$$

Hence, if at a large distance r the mass density is ρ, the filament sucks in matter at the rate

$$\frac{dM}{dt} = 2\pi r v_r \rho = (2\pi r)^2 \rho \frac{E_z}{\mu_0 I_z} \tag{17}$$

per unit length. If the electric field is antiparallel to the current, the drift is directed outwards.

Earlier, we applied these results to the solar corona. Prominences and spicules are likely to be of the same character.

Interstellar clouds often show a filamentary structure, indicating that currents flow along the magnetic field lines. The application of the contrast enhancement technique to photographs of interstellar nebulae demonstrates this in a dramatic way (Figure II.9). A cloud which in a usual photograph seems to have large, rather homogeneous regions, appears as a network of filaments. From what we have found in II.4, this should be interpreted as indicating filamentary and/or sheet currents. If this case is typical, we can conclude that homogeneous models of the formation and evolution of cosmic clouds are obsolete. Also, in this case 'homogeneity' is a consequence of inadequate information.

IV.8.3. DO MAGNETIC FIELDS AID OR COUNTERACT A COMPRESSION?

If the magnetic field is force-free it neither aids nor counteracts a contraction. The condition for this is that the electric currents producing the field flow parallel to the magnetic field. For a cylindrical geometry the field lines, as well as the current lines, are spirals of the type shown in Figure IV.7(b). If we change the angle between the current lines and the magnetic field lines from zero as in the force-free case to a finite value so that the current (i_1 in Figure IV.8) becomes more toroidal than the magnetic field, the electromagnetic pressure acts in the outward radial directions and counteracts a contraction. If, on the other hand, the current (i_2 in Figure IV.8) becomes more axial than the magnetic field, the electromagnetic forces tend to pinch the plasma.

Hence, decisive for the question of whether a magnetic field of cylindrical symmetry compresses or dilutes a plasma, is whether or not the magnetic field is more toroidal than the electric current.

We have already seen a number of examples of this in the magnetosphere, in the solar atmosphere, and in the heliosphere. In this respect the extrapolation of the heliospheric current model to galaxies is of special interest.

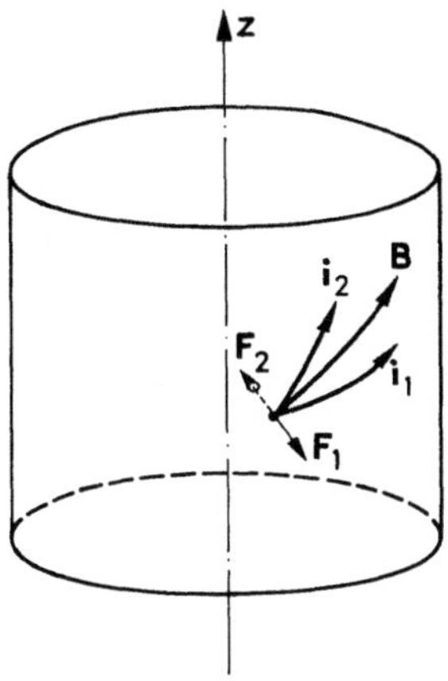

Fig. IV.8. Electric current–magnetic field configuration. If the current ($\mathbf{i}_1$) is more toroidal than the magnetic field, the electromagnetic force $\mathbf{F}_1 = \mathbf{i}_1 \times \mathbf{B}$ is directed radially outwards. If the current ($\mathbf{i}_2$) is more axial than the magnetic field, the force $\mathbf{F}_2 = \mathbf{i}_2 \times \mathbf{B}$ is directed radially inwards.

If we take $B_G = 10^{-9}$–10^{-10} T as a typical galactic magnetic field and $l_G = 10^{20}$–10^{21} m as a typical galactic length, the galactic current should be of the order $I_G = B_G \cdot l_G \cdot 2\pi/\mu_0 = 10^{17}$–$10^{19}$ A (Alfvén, 1977). In analogy with the heliospheric current system, this current might flow in the plane of symmetry of the galaxy, possibly waving up and down. It may consist of a number of more or less confined current regions. It is possible that part of this current flows through interstellar clouds and has assisted their formation.

IV.8.4. PINCH COMPRESSION OF DARK INTERSTELLAR CLOUDS

The possible importance of the pinch effect on interstellar clouds is illustrated by two examples.

(1) Consider first an interstellar cloud of a hundred solar masses, $M_G = 2 \times 10^{32}$ kg, occupying a volume of the linear dimension $l_C = 10^{17}$ m. The temperature of the cloud is $T_C = 10$–10^2 K. This represents a cloud of approximately the same mass as the Orion nebula. The number of atoms present in the cloud is $\simeq 10^{59}$, implying a mean density of $\bar{n} = 10^8$ m^{-3} and giving $N = 10^{42}$ atoms m^{-1}. Putting this latter figure and the temperature above into the Bennett relation (15) we find that an electric current of $I_C \simeq 5 \times 10^{13}$–$2 \times 10^{14}$ A has to flow through the cloud in order to produce a considerable compressional effect.

(2) Consider next a cloud of only one solar mass $M_C = 2 \times 10^{30}$ kg, having a temperature of $T_C = 10$–10^2 K and an extent of $l_C = 10^{16}$–10^{17} m. Hence, $\bar{n} = 10^6$–10^9 m^{-3} and $N = 10^{40}$–10^{41} atoms m^{-1}.

The current needed to compress the cloud is now found to be $I_C = 5 \times 10^{12}$–5×10^{13} A.

The electric currents obtained in the above two examples of interstellar clouds seem to be a fair geometric average between the heliospheric current of 3×10^9 A (10^8 e.m.u.) and the galactic current estimated at 3×10^{17} A (III.4.2 and III.4.4).

Further, the magnetic field in the cosmic cloud should be of the order of $B_C = \mu_0 I_c/2\pi l_c$ which gives $B_C = 10^{-9}$–10^{-11} T, also an acceptable range of values.

In the two examples above, the interstellar clouds were assumed to be contracted by a Bennett pinch, implying a pure toroidal magnetic field and a pure axial current. According to the discussion in IV.8.3, all magnetic configurations ranging from this kind of pinch to the force-free states would also give rise to contractive forces. For the plasma to pinch in these cases the total current must be larger than that given by the Bennett relation

In conclusion, it can be stated that the pinch mechanics can compress interstellar plasmas into dense clouds. This is important to the theory of interstellar clouds, because gravitation need not necessarily be postulated as the only force which generates them and dominates their further evolution. *The formation and evolution of interstellar clouds may be controlled, or rather is likely to be controlled by electromagnetic effects.*

IV.9. Ambiplasma

IV.9.1. PROPERTIES OF AN AMBIPLASMA

As we shall see in Chapter VI, there are now rather strong arguments for the existence of antimatter. Hence it is important to study the properties of a plasma consisting of a mixture of koinomatter* and antimatter. Such a plasma is usually called an *ambiplasma* (*ambi* = both).

In general, an ambiplasma may contain a mixture of all elements and antielements. In order to simplify the treatment we shall confine ourselves to the case when it has only four constituents, n_p^+ protons, n_p^- anti-protons, n_e^+ positrons, and n_e^- electrons per unit volume. As all cosmic plasmas, it is magnetized (field = B), and it may be acted upon by gravitation g (which of course acts in the same way on particles and antiparticles).

The quasineutrality condition requires

$$n_p^+ + n_e^+ = n_p^- + n_e^-. \tag{18}$$

There are three important special cases:

Koinomatter:

$$n_p^- = 0, \quad n_e^+ = 0, \quad n_p^+ = n_e^-; \tag{19}$$

Antimatter:

$$n_p^+ = 0, \quad n_e^- = 0, \quad n_p^- = n_e^+; \tag{20}$$

Symmetric ambiplasma:

$$n_p^+ = n_p^-, \qquad n_e^+ = n_e^-. \tag{21}$$

IV.9.2. THE ANNIHILATION REACTIONS

A proton and an antiproton may annihilate each other. The annihilation results in neutrinos, and high-energy γ rays. After the π and μ decays there are, on the average,

* Koinomatter (from Greek *koinos* = common) designates ordinary matter. Both antimatter and koinomatter are of course *matter*, have mass and are subject to gravitation in the same way.

about three electrons plus positrons per annihilation, and their energy is of the order 100 MeV. The energy spectrum is shown in Figure IV.9.

The processes of annihilation depend on the initial energy. The p^+p^- annihilation reaction is dominated by the strong interaction and produces principally positive, negative and neutral pions, which very rapidly decay into electrons, positrons, neutrinos and photons. The principal e^+e^- reaction that leads to direct annihilation is the electromagnetic interaction in which two photons are produced.

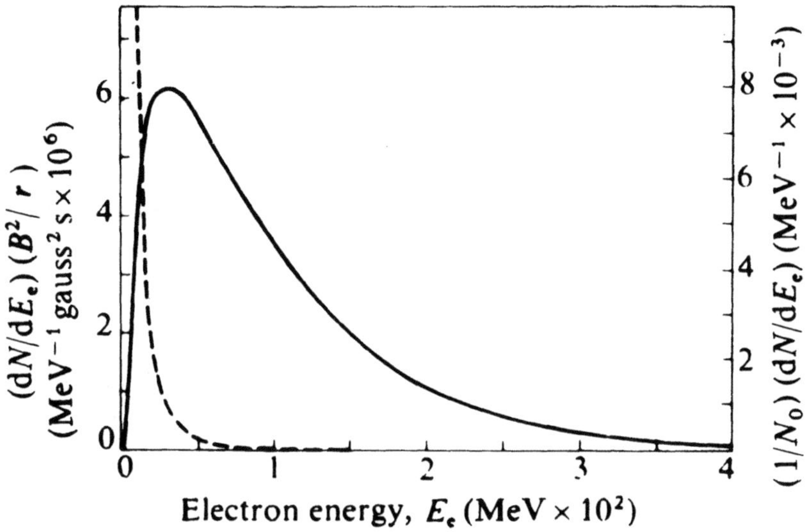

Fig. IV.9. ——, Energy spectrum of electrons (negative and positive) directly produced by proton-antiproton annihilation. – – –, Stationary energy spectrum of annihilation electrons dN/dE_e obtained for a constant rate of annihilation when the electrons lose energy to synchrotron radiation in a magnetic field **B**.

Experiments with antiprotons have been made at megavolt and gigavolt energies, but in many of our problems, thermal energies are relevant, and we do not have any measurements of the annihilation cross-section for either the proton–antiproton or the electron–positron reactions. We have to use theoretical estimates which necessarily are uncertain.

Introducing the 'classical radius of the electron'

$$d = \frac{\mu_0}{4\pi} \frac{e^2}{m} = 2.8 \times 10^{-15} \text{ m} \tag{22}$$

and the cross-section

$$\sigma_0 = \pi d^2 = 2.5 \times 10^{-29} \text{ m}^2 \tag{23}$$

we can write the life-time of an electron (positron) in a positron (electron) gas with number density n_e, as

$$t_e = k_e t_{0e} \tag{24}$$

with

$$t_{0e} = (n_e c\sigma_0)^{-1} = 4.2 \times 10^{12}\, n_e^{-1} \text{ years.} \tag{25}$$

For protons and antiprotons it will be

$$t_p = k_p\, t_{0p} \tag{26}$$

with

$$t_{0p} = (n_p c\sigma_0)^{-1} = 4.2 \times 10^{12}\, n_p^{-1} \text{ years.} \tag{27}$$

The energy dependency is now represented by the factors k_e and k_p. Their values must, for the reasons mentioned above, be estimates. According to Morgan and Hughes (1970) they can be treated as different functions in different kinetic energy ranges. Thus for highly relativistic electrons the value follows

$$k_e = \frac{\text{eV}}{mc^2} \ln \frac{\text{eV}}{mc^2}. \tag{28}$$

In the range 10 keV to 1 MeV, k_e is close to unity. For energies below 10 keV, however, the effect of Coulomb attraction must be accounted for. Therefore, the value should be put

$$k_e = \frac{v}{c} \frac{(1 - e^{-2\pi\alpha c/v})}{2\pi\alpha} \tag{29}$$

with the fine structure constant $\alpha = 1/137$. The value of $k_p(v)$ is approximately $5k_e(v)$ for energies below 1 GeV. Thus, between 10 MeV and 1 GeV its value is 5, and below 10 MeV

$$k_p = \frac{5v}{c} \frac{(1 - e^{-2\pi\alpha c/v})}{2\pi\alpha}. \tag{30}$$

These estimates pertain to the direct annihilation process. However, under certain conditions other interactions might precede the annihilation, thus considerably changing the effective cross-sections.

For low energies e^-e^+ may undergo radiative capture to form a bound state of the positronium which later will annihilate. This process increases the annihilation cross-section, sometimes by several orders of magnitude. Because the radiative processes are less efficient for p^+p^- radiative capture is insignificant for them.

In a hydrogen–antihydrogen mixture containing atoms and molecules, various rearrangement collisions occur that greatly influence the e^+e^- and the p^+p^- annihilation rates.

The cross-sections and lifetimes for both electrons–positrons and protons–antiprotons are found in the tables and diagrams of the quoted paper.

IV.9.3. RADIATIONS FROM AN AMBIPLASMA

The electrons and positrons, which are the result of the proton–antiproton annihilation have energies of the order 10^8 eV. All known cosmic plasmas are magnetized, and there is no reason to suppose that the case treated here should be different. In a magnetic field B the electrons and positrons spiral and emit synchrotron radiation. The decay time is

$$t_s = \frac{5}{B^2} \frac{1}{1 + (W/W_0)} \text{ s} \tag{31}$$

(see, for example, CE, p. 69), where W is the kinetic energy and $W_0 = m_e c^2$. If $W = 100$ MeV $= 200\,W_0$, we obtain

$$t_s = 2.5 \times 10^{-2} B^{-2}. \tag{32}$$

If $t_s \ll t_e$ the electrons and positrons radiate most of their energy as synchrotron radiation before they annihilate. If, on the other hand, $t_s \gg t_e$, very little synchrotron radiation is emitted.

The energy maximum of the spectrum of the synchrotron radiation is located at the frequency

$$\nu_{\max} = \frac{eB}{2\pi m}(W/W_0)^2 = 2.8 \times 10^{10}(W/W_0)^2 B. \tag{33}$$

With $W/W_0 = 200$, we have

$$\nu_{\max} = 10^{15} B. \tag{34}$$

It should be noted that if the particles spend most of their time in a field B_0, but 1% of their time in a field $B_1 = 10B_0$, half of their energy is radiated in the stronger field. Hence, both for the calculation of the lifetime and the maximum of the spectrum one should rather use the maximum field than the average field.

Since real cosmic plasmas are inhomogeneous, this is important. The emitted synchrotron radiation may be much more intense than if calculated from the average value of B.

Compared to the synchrotron radiation, the Bremsstrahlung is negligible in space, but may be of decisive importance in regions where the density is high.

Under the assumption that $t_s \ll t_p$, the net result of a proton–antiproton annihilation is the emission of neutrinos of about 1000 MeV, γ-rays totalling about 700 MeV, and the remaining 300 MeV in the form of e^-e^+-pairs. Thus, these fractions of the released annihilation energy are one half, one third and one sixth, respectively.

The positrons and electrons also annihilate with the emission of photons. The photon energy is 0.5 MeV *in the rest frame*. However, if t_s is not small enough they will annihilate before they have lost all their kinetic energy. This leads to the observed γ-rays having any energy between zero and a maximum depending on the kinetic energies. The result is that a wide spectral band is emitted. Therefore, *the existence of antimatter does not necessarily manifest itself by the emission of the* 0.5 MeV *line*.

In addition, an ambiplasma emits neutrinos, but they are impossible to detect with present techniques. The ambiplasma may also emit light and all other types of radiation which characterize a cosmic plasma.

In an infinite plane layer of thickness Δ the total energy per m^2 is

$$W_M = 2m_p c^2 n_p \Delta \tag{35}$$

where n_p is the number of p^+p^- per unit volume.

This energy is radiated during a time t_p given by Equation (26). The energy flux ϕ per unit time is

$$\phi = W_M/t_p = 2m_p c^3 \sigma_0 n_p^2 \Delta k_p^{-1} \tag{36}$$

$$\phi = 2.2 \times 10^{-30} n_p^2 \Delta k_p^{-1}\ \mathrm{J\,m^{-2}\,s^{-1}}. \tag{37}$$

IV.9.4. MAIN AMBIPLASMA PROBLEMS

There are three main problems with cosmic ambiplasma: (i) co-existence of matter and antimatter, (ii) annihilation as an energy source, and (iii) separation of matter and antimatter.

(i) *Co-existence of koino-matter and antimatter.* Is it possible for matter and antimatter to co-exist in the universe? If so, could they co-exist even within our galaxy (see VI.2).

As long as all astrophysical models were basically homogeneous, the answer had to be negative. The modern views according to which plasmas usually are inhomogeneous and especially the discovery of the cellular structure of space has changed the situation.

(ii) *Annihilation as an energy source.* Annihilation seems to be the only reasonable source of energy for a number of extremely energetic phenomena, such as the Hubble expansion, the radiation from QSO's, and a number of other phenomena which are now being explored by X-ray and γ-ray astronomers (IV.10).

(iii) *Separation of koino-matter and antimatter.* If we accept Klein's model which asserts that the universe 'originally' consisted of a very tenuous ambiplasma, processes separating koino-matter and antimatter are of decisive importance. However, even if we accept the view that the present universe is symmetric, we need not necessarily accept the Klein version of a symmetric cosmology. More about this in Chapter VI.

IV.9.5. SEPARATION OF MATTER AND ANTIMATTER

Gravitation alone cannot separate an initially symmetric ambiplasma (containing equal quantities of matter and antimatter) into matter and antimatter. However, in principle, this is possible under the action of electromagnetic effects (Alfvén and Klein, 1963).

In order to demonstrate this let us assume that a symmetric ambiplasma is located in a gravitational field which concentrates the protons and antiprotons in a lower region and the electrons and positrons at higher levels. Suppose that there is an electric current parallel to g. This will remove negative charge from the top of the atmosphere, and since the highest region contains mainly positrons and electrons, it is essentially the electrons that are removed. Simultaneously, the current will remove negative charge from the base of the atmosphere, and since the lowest region contains mainly heavy particles, the result will be essentially a loss of protons. Hence the current will concentrate antimatter at the interphase between the p^+p^- and the e^-e^+ regions. Similarly, a vertical current antiparallel to g will cause a concentration of matter at the interphase. Hence a current in the circuit shown in Figure IV.10 will produce one region of matter and another of antimatter.

It may be objected that a process of this kind is in contradiction to the general laws of statistical mechanics, because it changes a state of disorder – matter and antimatter homogeneously mixed – into a state of higher order. This objection is not valid, because the ultimate source of energy is the gravitation of the whole system. Consider as an analogy water pouring down from a mountain. It is possible to convert its gravitational energy into electric energy by means of a hydroelectric plant, and the electric energy can be used to electrolyze part of the water and separate it into hydrogen and oxygen.

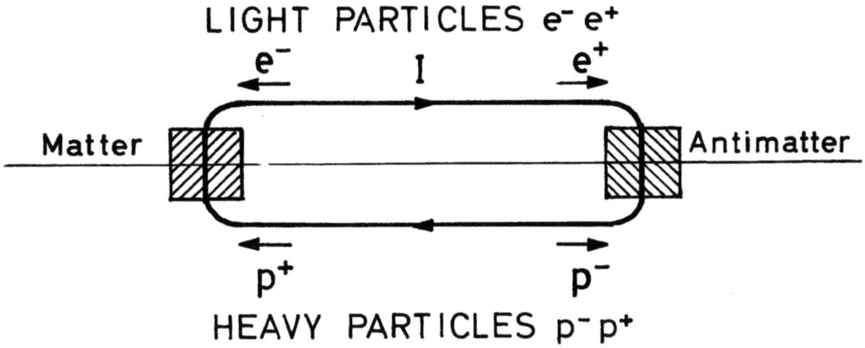

Fig. IV.10. An ambiplasma in a gravitational field will be separated by an electric current, so that matter will be concentrated where the current is antiparallel to g, and antimatter where the current is parallel to g.

Our process is similar. In fact, the separation of matter or antimatter out of a symmetric ambiplasma can be considered as a result of electrolysis. The quantity m of koino-matter or antimatter which is separated by a current I flowing during a time t is

$$m = (m_{\mathrm{H}}/e)\,It \tag{38}$$

(where m_H is the mass of the hydrogen atom) in agreement with the ordinary law of electrolysis.

As currents are normally produced in a moving magnetized plasma, a process of this kind should occur as a natural consequence. However, the total quantity which can be separated in a certain closed-current circuit is limited. Suppose that a current $I = \pi r^2 i$ flows along the axis of a circular cylinder with radius r. The magnetic field at the surface of the cylinder is

$$B = \frac{\mu_0 I}{2\pi r}. \tag{39}$$

Combining this equation with Equation (38), we obtain

$$\begin{aligned} m &= m_H t\, 2\pi r\, B/\mu_0 e \\ &= 5 \times 10^{-2}\, trB. \end{aligned} \tag{40}$$

The magnetic fields in space are not usually supposed to exceed 10^{-9} T, or possibly 10^{-8} T. As the phenomena we discuss take place in regions where the energy density is large, the magnetic fields may also be large. The time t cannot reasonably exceed 10^{17} s ($= 3 \times 10^9$ years). With $B < 10^{-8}$ T, and $t < 10^{17}$ s we obtain

$$m < 5 \times 10^7\, r. \tag{41}$$

Hence, a separation on a galactic scale ($r = 10^{21}$ m) would not suffice to separate one solar mass.

The process is much more efficient on a small scale. If the medium has a density nm_p the mass is $nm_p r^3$. Hence a total separation can be obtained if

$$nm_p r^3 = 5 \times 10^{-2}\, trB \tag{42}$$

or

$$r^2 = 5 \times 10^{-2}\, tB/m_p n = 3 \times 10^{25}\, Bt/n. \tag{43}$$

The time t is limited by annihilation. Introducing t_0 from Equation (27) we obtain

$$r = 6 \times 10^{22} B^{1/2} n^{-1}. \tag{44}$$

With $B > 10^{-4}$ T and $n = 10^2$ m^{-3} we obtain $r < 10^{19}$ m and $m < m_p n\,(10^{19})^3 < 10^{31}$ kg.

Hence a 'small scale' structure with elements whose sizes are 0.1 light year or less may produce a high degree of separation between matter and antimatter. The mass separated by each element may not be very large, but the total mass may be a large fraction of the whole ambiplasma.

A large number of similar plasma processes have been suggested which separate matter and antimatter (see Thompson, 1978). It has been suggested by Thompson that in general, an ambiplasma represents an unstable situation and that it spontaneously 'coagulates' into regions of matter and antimatter. By such processes and/or subsequent coalescence of regions containing the same kind of matter, Thompson claims that a separation of galactic scale can take place.

IV.9.6. ON THE CO-EXISTENCE OF MATTER AND ANTIMATTER

The average density in the metagalaxy is estimated to be 1 particle m^{-3}. According to what is said above, a homogeneous ambiplasma of this density has a very large lifetime. Within a galaxy the average density is of the order of 10^6 m^{-3} particles. This means that if our galaxy consisted of homogeneous ambiplasma, it would have annihilated long ago. Antimatter can exist in a galaxy only if it is separated from matter. We discuss here the interface between matter and antimatter.

If, as an illustration, a drop of water is placed on a hot-plate with a temperature somewhat above 100°C, it evaporates very rapidly. However, if the hot-plate is very hot the water may remain for five or ten minutes (especially if the plate is a little concave). The evaporation of the water produces a thin layer of vapor which isolates the water from the hot-plate. In physics textbooks the phenomenon is referred to as the 'Leidenfrost phenomenon'. In high temperature boilers it is of technical importance.

If matter and antimatter are brought together, an analogous phenomenon may occur. At the interface between matter and antimatter, annihilation produces a hot layer which separates matter from antimatter in the same way as the water vapor in the Leidenfrost layer separates the water from the hot-plate.

We shall discuss this further in Chapter VI.

IV.10. High Energy Phenomena

An important and puzzling problem in astrophysics is how the energetic, sometimes extremely energetic radiation is produced and how matter is accelerated to velocities sometimes approaching the velocity of light. Half a century ago it was discovered that the Earth receives an (almost) isotropic radiation consisting mainly of protons and other nuclei, with very large energies. This radiation is referred to as cosmic radiation, nowadays often 'galactic' cosmic radiation although it has not been proved that the radiation we receive on Earth is necessarily generated in the galaxy. Particle energies up to 10^{11} eV have been measured directly, and from the study of large showers, it is concluded that some primary particles have energies of at least 10^{19} eV.

Another high energy radiation called solar cosmic rays emanates from solar flares. Particle energies reach 10^9, or sometimes 10^{10} eV.

High energy particles, both ions and electrons, are also observed in the magnetospheres. The first to be discovered was the Van Allen radiation, and similar radiation belts are known to surround Jupiter and Saturn. The energy reaches 10^6–10^8 eV. Further, in the auroral zone the ionosphere is hit by beams of electrons with energies in the 10^3–10^4 eV range.

The advance of X-ray and γ-ray astonomy has demonstrated the existence of a large number of astrophysical objects emitting high energy photons. Some radiation, especially the ordinary 'galactic' cosmic rays are time-constant, or show slow time variations. In other cases, the radiation varies rapidly, with time constants down to fractions of a second (γ-ray burst) (see VI.3.6).

Furthermore, distant galaxies have redshifts, indicating that they move away from us with kinetic energies of up to 10^8 eV proton^{-1}. Some QSO's have even larger redshifts, indicating kinetic energies of 10^9 eV proton^{-1}. The total kinetic energies of these objects are, of course, very much larger than the energy of the radiation.

IV.10.1. SOURCES OF ENERGY AND ACCELERATION PROCESSES

There are a number of theories attributing some of the observed radiation to special mechanisms deriving their energy from double stars, novae, supernovae, pulsars, etc. The enormous kinetic energies of galaxies and QSO's are usually attributed to phenomena at the singular point of the big bang hypothesis, or rather, are *postulated* as initial conditions at the assumed creation of the universe.

In the following we shall investigate here to what extent it is possible to avoid such *ad hoc* assumptions, and regard the high energies as resulting from general plasma phenomena which we have reason to believe must exist without making any *ad hoc* assumptions.

We shall now discuss some important acceleration mechanisms.

IV.10.1.1. *Varying Magnetic Fields*

Time variations of magnetic fields pump the energy of a particle up and down. If particles are scattered by magnetic field irregularities, the pumping results in a net energy gain. This can accelerate the particles to any energy which allows the particles to be confined by the magnetic field. The Fermi process can be considered as a special case of this. In addition, the acceleration of charged particles by different types of waves in a plasma is also related to this. These acceleration mechanisms should fill certain regions of space with high energy particles. The observed net effect is a particle radiation with slow intensity variations and a high degree of isotropy. See further IV.10.2.

IV.10.1.2. *Acceleration in Double Layers*

When charged particles pass a double layer with a voltage drop ΔV_0 they pick up an energy $ze\Delta V_0$ and primarily form a monochromatic beam. If a double layer explodes so that ΔV suddenly increases by orders of magnitude to ΔV_{max}, the particle energies will cover a spectrum from $ze\Delta V_0$ up to $ze\Delta V_{\mathrm{max}}$. However, as an exploding double layer produces violent disturbances in the surroundings, there will be a number of complicated secondary processes which transfer the primary energy to other particles and degrade it. At the same time, the highly irregular and rapidly fluctuating magnetic fields which are produced will pump some particles to energies which are even higher than the $e\Delta V_{max}$ delivered by the double layer.

The emitted radiation will show very rapid and erratic fluctuations. However, if the particles are stored in large volumes for a long time, these variations may be smoothed.

IV.10.1.3. *Annihilation*

As we shall discuss in more detail in Chapter VI, there are reasons to suppose that the universe is matter–antimatter symmetric. This means that in certain regions where the koino-matter and antimatter interact, there should be intense generation of energetic radiation. From particle physics studies in the laboratory, we can predict some of the effects, but the behavior of an ambiplasma is necessarily hypothetical and speculative (IV.9).

Proton–antiproton annihilation produces mesons of a total energy of 2×10^9 eV and at their decay γ-radiation is produced with a spectrum up to 10^8 or 10^9 eV. Also, electrons–positrons with similar energies are produced.

When degraded, the high energy radiation gives rise to a number of particles (including photons) with lower energies. Again there will be violent irregular motions in the surrounding plasma, and magnetic pumping as well as related processes may accelerate a few particles to energies much higher than the primary energies.

The total energy stored as kinetic energy in galaxies taking part in the Hubble expansion represents the largest energy in the universe with the exception of the rest mass. If we want to explain this according to normal scientific procedures, i.e., without introducing new laws of physics or supernatural phenomena, the only possible source of energy is annihilation; However, the usual 'explanation' is to attribute it to processes in the singular point in the big bang hypothesis. It is often claimed that it is the result of an 'explosion' at or near the singular point, but as the Friedman model is isotropic, there can be no pressure gradient to cause the acceleration. In reality, the big bang theory *postulates* the existence of the high velocities in the singular point without any attempt to account for them as produced by any physical mechanism. This means that when an observed redshift is said to be of cosmological origin, the word 'cosmological' is used as an euphemism for 'supernaturally produced'.

IV.10.1.4. *Gravitation*

Gravitation is, of course, one of the dominating forces in astrophysics. However, as electromagnetic forces are stronger by a factor of 10^{39}, gravitation is important only when electromagnetic forces neutralize each other, as is the case for large bodies. In our solar system, gravitational forces do not seem to be of primary importance in producing high energy phenomena.

A solid body (e.g., meteroid, comet or asteroid) hitting the Earth with escape velocity has an energy of only 1 eV nucleon^{-1}. For the Sun, the value is 10^3 eV nucleon^{-1}. In the collision, most of the energy is dissipated as heat and low-energy photon radiation, so that higher energies are obtained only through complicated and inefficient processes.

Diffuse media, such as gas or plasma clouds, falling in towards celestial bodies, are usually still more inefficient. A plasma cloud approaching the Earth will already be stopped in the magnetosphere, or in any case, in the upper ionosphere, where gas clouds will also be stopped. The result will be a heating of the upper atmosphere which makes it expand and stop an additional infall of more distant clouds.

Low density plasma clouds approaching the Sun will be stopped very far away by the solar wind. A neutral gas cloud falling towards the Sun is likely to be stopped when it has reached the critical velocity (see Chapter V). This occurs when the cloud is still very far from the photosphere. In fact, its kinetic energy will be only of the order of 10 eV when this occurs.

Even if gravitational energy is an unimportant source of high energy acceleration in our neighborhood, we cannot exclude that it may be of importance in stars, e.g., pulsars, with large escape velocities. However, falling, solid bodies will dissipate their energy through inelastic collisions, and only a small amount of their energy can be expected to be so concentrated that it is likely to be converted in any high energy phenomena. Diffuse clouds should also be stopped by the same type of processes as occur in the solar system. Hence we conclude that although we cannot rule out gravitation as a source of high-energy particles, it is probably not a very important one.

It has been claimed that there exist 'black holes' which convert gravitational energy into different kinds of radiation. This cannot be excluded.

There is no doubt a need for an energy conversion process which can explain the large amounts of energy appearing in the form of different kinds of energetic radiation, the Hubble velocities and the still greater velocities of QSO's. The only sources with a capacity large enough are gravitation and annihilation. The arguments in favour of such processes connected with black holes are, however, not very convincing. The same can be said even for the arguments supporting the existence of black holes. An unbiassed comparison should be made between the black hole and the annihilation approaches.

IV.10.2. MAGNETIC PUMPING

As double layer acceleration is discussed in Chapter II, and annihilation in Chapter VI, we shall confine ourselves here to magnetic pumping.

The acceleration of particles to high energies by magnetic field variations has attracted much interest. The pioneer investigator was Swann, who pointed out that in an increasing magnetic field, e.g., in a sunspot, charged particles could be accelerated to extremely high energies. Vallarta called this process the 'Cygnotron' acceleration (cygnus-swan) but when Kerst invented an accelerator based to some extent on a similar process, his term 'betatron' became generally accepted.

Suppose that particles spiral in a homogeneous magnetic field having the energies $W = W_\perp + W_\parallel$ (kinetic energy perpendicular and parallel to the magnetic field B_0). If the magnetic field increases to B_1, $W_\perp$ will increase but $W_\parallel$ remains unchanged. If later the magnetic field decreases again to B_0 the particles will be decelerated and their net gain in energy is zero.

If, however, the particle is scattered, e.g., by magnetic field irregularities, when the magnetic field is B_1, part of the energy gain may be transferred to $W_\parallel$ and when the magnetic field returns to B_0 this will be conserved. Later it may in part be transferred to $W_\perp$ by another scattering process. The maximum energy which a single particle trapped in a certain magnetized region can reach by iterative pumping is limited only by the magnetic field. When the particle energy becomes too large, the field is not strong enough to keep it trapped. The case when the magnetic field varies as a sine function has been studied by Schlüter (1957). For situations where the distribution of momentum is not too far from isotropic, Schlüter has shown that

$$\frac{\mathrm{d}}{\mathrm{d}t}\frac{W_\perp^2 W_\parallel}{B^2} = \frac{\nu}{3}\frac{W_\perp(2W_\parallel - W_\perp)^2}{B^2} \tag{45}$$

where $W_\parallel = p_\parallel^2/2m$, and ν is the collision frequency.

Magnetic pumping is used for heating laboratory plasmas. In space, the requirement for this process to work is that the magnetic field varies, and that there are magnetic irregularities scattering the particles, which is also a realistic assumption (in order to make the process rapid, the time constants for the scattering and for the magnetic variations should be of the same order of magnitude).

If the particles are confined by two magnetic mirrors which approach each other, a similar acceleration may take place. This is the well-known Fermi process for accelerating cosmic rays. Somewhat similar ways of accelerating particles to cosmic ray energies have been discovered, for example, by Van Allen.

If a large number of particles are accelerated, the maximum energy they reach is given by the equipartition condition: the particle energy per unit volume cannot exceed the magnetic energy fluctuations or the plasma kinetic energy.

Magnetic pumping gives an energy spectrum which is reconcilable with the observed spectrum and the mechanism also seems to explain other observed properties (Fälthammar, 1963).

IV.10.3. REGIONS OF COSMIC RAY ACCELERATION

If we accept that magnetic pumping may explain the observed cosmic radiation as a general property of a space plasma, then the next step is to determine where the acceleration which produces the cosmic radiation we observe takes place. The answer to this question seems to be straight-forward. Since the radiation is generally called 'galactic' cosmic rays, it must obviously be produced in our galaxy. However, scientific questions cannot be solved by inventing a term and then getting it 'generally accepted'. Surely, if we analyse the motivation for calling the radiation 'galactic', we find that the only reason for the name is the existence of certain models which one day are likely to share the fate of so many other astrophysical models.

In the following sections, we shall discuss an alternative approach to cosmic ray problems. As this is based on assumed phenomena, which take place in regions which are not accessible to *in situ* measurements, this approach, like all of the others, is necessarily speculative.

IV.10.3.1. *Heliospheric and Galactic Cosmic Radiation*

We know that in the Earth's magnetosphere, a high-energy particle radiation is generated. The same is true for the magnetospheres of Jupiter and Saturn. There seems to be no reason why it should not also be true for the heliosphere. It is easily seen that the emitted solar wind energy is enough to accelerate cosmic rays, in addition to covering the expected energy losses due to absorption in heliospheric matter. Hence, the only question is how high the cosmic ray energies are, which can be kept trapped by the heliospheric magnetic field. Since the magnetic field in the outer heliosphere or transplanetary space is not accessible to spacecraft measurements, we know nothing with certainty about this. Extrapolation from the known parts of the heliospheric magnetic field could be done in different ways and will, in any case, be necessarily speculative. Order of magnitude estimates could be made in several ways: the limit to the heliosphere is certainly at least $R = 10^{12}$ m. Putting $B = 10^{-9}$ T, we have for the order of magnitude $W_h = cB\rho = 3 \times 10^{11}$ eV. However, if we include the transplanetary region which may also be magnetized, the value may very well be increased to $R = 10^{14}$ or 10^{15} m and $W = 3 \times 10^{13}$ or 3×10^{14} eV.

In the following we shall tentatively use $W_h = 10^{13 \pm 1}$ eV for the maximum energy which can be trapped in the heliospheric magnetic field (this may include a large transplanetary region). It is generally supposed that the interplanetary magnetic field connects directly to a homogeneous interstellar field in a way which allows free passage of cosmic rays. With our general hesitation to accept homogeneous models, there is no reason why we should accept this. Rather, we should consider the heliospheric magnetic field to have the same property as the fields of the Earth, Jupiter and Saturn, viz. to be able to confine the particle radiation it generates up to the limit which is set by its rigidity ($B\rho$ value). The tentative conclusion from this is that what we may define as the medium energy cosmic ray, up to $W_h = 10^{13 \pm 1}$ eV, is generated and trapped in the heliosphere, and should be called '*heliospheric cosmic ray*', It should, of course, not be confused with 'solar cosmic rays' generated by solar flares.

There is no reason to suppose that in the energy range $< W_h$ the intensity outside the heliosphere is the same as measured here, as is claimed by a number of speculative theories. Indeed it may very well be much smaller, perhaps orders of magnitude smaller.

However, cosmic rays above W_h cannot be trapped in the heliospheric magnetic field, and so will necessarily be able to move out in interstellar space. Hence for $W > W_h$ measurements on Earth give us reliable information about the intensity and spectrum in interstellar space, but only in a region comparable to the Larmor radius of the radiation.

The next question is whether the high energy cosmic rays ($W > W_h$) can be trapped in our galaxy. If we tentatively put $\rho = 10^{20}$ m and $B = 10^{-10}$ T as characteristic values for our galaxy we find $W = 3 \times 10^{18}$ eV. Since this is not far from the upper limit for cosmic rays (allowing for the uncertainties) we may conclude that there is no decisive reason why cosmic radiation above W_h could not be trapped in our galaxy. So it would be appropriate to refer to the cosmic rays that we observe in the energy range above W_h as *galactic cosmic rays* (subject to future revision, if we find decisive arguments for assuming them to move out in intergalactic space). It should be noted that the total energy of cosmic rays observed above W_h is only a small fraction of the heliospheric cosmic radiation.

IV.10.3.2. *Origin of Galactic Cosmic Rays*

Of the three mechanisms producing high energy phenomena, we can exclude annihilation as a source of cosmic rays above 10^9 eV. Magnetic pumping is able to produce the heliospheric cosmic radiation, and could very well also produce galactic cosmic rays by acceleration in interstellar space up to energies far above W_h. However, as we have seen in II.6.3–II.6.5, double layers may also produce extremely high energies. This is known to take place in solar flares, where they generate solar cosmic rays of up to 10^9–10^{10} eV. In III.4.4, a mechanism in double radio sources which may accelerate particles to cosmic ray energies was discussed. Whether we should allow ourselves to extrapolate this mechanism up to 10^{19} eV is an open question.

CHAPTER V

ORIGIN OF THE SOLAR SYSTEM

V.1. How We Can Reconstruct Earlier Epochs

In the preceding chapters we have seen how drastically *in situ* measurements in space have changed our views of the properties of cosmic plasmas. It has also been pointed out that these changes are not likely to be confined to the region where the measurements have been made. A plasma has basically the same properties in the magnetospheres as in the laboratory, and it is also likely that the plasmas in regions which are remote in space or in time have basically the same properties as in the present magnetospheres. Hence the conditions at earlier epochs should be treated as extrapolations of the present-day conditions in the magnetospheres (including the heliosphere) and in the upper ionospheres.

Assuming that the Sun and the solar system formed out of a dusty cloud, and paying special attention to plasma effects (Alfvén and Carlqvist, 1978), we arrive at a model for the evolution of the solar system which is reconcilable with a model derived from present solar system data exclusively (see ESS). In this way we seem to reach a consistent picture of the evolutionary chain from the formation and evolution of interstellar clouds to the present state of the solar system. Of course, this picture is only preliminary because so many links in the chain are still necessarily uncertain, and need to be supplemented by new observational and theoretical data.

V.2. Sources of Information

Up until quite recently the only method of investigating the early solar system back to the formation of the celestial bodies around the Sun was to derive the evolutionary process from an analysis of the present data about the chemical, and especially the dynamic properties of these bodies. Space research has supplied us with much information in these respects, and combining this with what we know about the properties of plasmas from the laboratory, a reconstruction of the evolution back to the formation of the solar system has been made by Alfvén and Arrhenius (ESS). (Concerning other attempts to reconstruct the early solar systems, mostly without taking account of plasma effects, see, e.g., Safronov (1969), Reeves (1972) and Gehrels (1978).)

It is shown that at the time of formation and early evolution, plasma processes were decisive, but that their importance decreased when planetesimals were formed. Celestial mechanics dominated the later evolution. In the present epoch magnetohydrodynamic effects are not important for the motion of the big celestial bodies (planets, satellites and also asteroids) but are still influencing the evolution of comets and, of course, the state of the magnetospheres (including the heliosphere) and ionospheres.

Especially during the last few years infrared and radio astronomy have caused a rapid advance in our knowledge of interstellar clouds. As it is likely that the Sun and the solar system once formed out of such a cloud (presumably a dark cloud) this knowledge is now a valuable supplement. Such clouds are known to consist of a dusty plasma, which means that the properties of *dusty plasma clouds* are decisive for the earliest periods of the formation of the Sun and the solar system (IV.7 and IV.8).

V.3. Impact of Magnetospheric Results

When judging different approaches to the evolutionary processes, we should do so with what we have found in Chapters I–IV as background.

Magnetospheric research together with laboratory plasma investigations have given us a better understanding of some basic properties in diffuse media. We can now – to some extent – 'translate' the observed phenomena over large parameter ranges (see Figure I.3) so that we can predict the properties of plasmas from studies made at a different range of parameters. This means that we can transfer knowledge from magnetospheric and laboratory studies to investigations of, e.g., interstellar clouds.

The normal description of the hydromagnetic state of a space medium by magnetic field models can only account for some of the phenomena (see Figure I.4). It is necessary to translate magnetic field models into electric current models in order to understand a number of phenomena which are important for the properties and the evolution of the medium.

V.3.1. ACTIVE AND PASSIVE PLASMA REGIONS

Cosmic plasmas generally seem to consist of regions of active plasma and regions of passive plasma (II.9). In short:

(a) *active regions* carry field aligned currents and neutral line currents. These produce heating, transfer energy and often generate *double layers*, which accelerate particles to sometimes very high energies.

(b) in *passive regions*, currents along field lines are small. Passive regions can transmit waves and high energy particles, but these are essentially energized by phenomena in the active regions. Although often the active regions are small compared to the passive regions, they are decisive for the evolution of cosmic clouds.

The properties of these two regions are so different that any attempt to take an average is misleading. Since the active regions are often much smaller than the passive regions, averaging will easily lead to a neglect of the active regions. For example, a global average of properties of the ionosphere would easily lead to a neglect of the narrow auroral zones, and the crucial importance of them to many phenomena will not show up. Similarly, observations of distant objects will often only give us averages which may lead to a similar neglect of important mechanisms. The contrast enhanced photographs (see, e.g., Figure II.9) demonstrate a small-scale structure in clouds which earlier were believed to be fairly homogeneous. If such structures also exist in dark clouds (which seems reasonable from what we know of plasmas in the magnetosphere) this will be decisive for understanding the properties of these clouds and their possible evolution to solar systems.

V.3.2. EXTERNALLY DRIVEN CURRENTS

The properties of a cosmic cloud through which an electric current flows are not determined only by the parameters (like density, temperature, magnetization, state of ionisation) in the cloud, but depend in an often important way on the *whole circuit* in which the current flows. Hence, if the magnetisation of a cloud is non-curlfree (i.e., an electric current flows there) the properties of the cloud may be determined by processes taking place very far away. In other words, the boundary conditions are essential. If the current through the cloud is part of a general galactic current system (as it is likely to be in many cases) the cloud may be energized by an e.m.f. located in another part of the galaxy, and its evolution determined by the conditions far away.

Consider a cloud where a current, which is closed through a circuit much larger than the cloud, flows. (This current may have once induced the formation of the cloud.) The circuit elements are an e.m.f. V_b, which is produced, for example, by the conversion of kinetic energy into circuit energy, an inductance L and a resistance R_0 (see Figure II.16, where the cloud is represented by the discharge tube). We assume that L and R_0 are large enough to keep the current I in the circuit essentially constant. If in one part of such a circuit, for example in the cloud, the energy losses are increased, we would expect the plasma to be cooled. We know from laboratory investigations that the result is the opposite. Since the cooling causes the electric field E driving the current to increase, the power dissipation in the cloud *increases* so much that the electron temperature increases. This is necessary in order to compensate for the increased losses, and the power is taken from the whole circuit (primarily from the magnetic energy stored in L).

There seems to be a widespread belief that if in the evolution of a cloud its temperature falls, and its density increases, *magnetic fields may be expelled* because of a decrease in conductivity. If we discuss such statements from the electric current point of view, it is obvious that a magnetic field cannot be 'expelled' unless the current pattern changes. Further, the current flows mainly in the active plasma regions, so the current pattern may remain essentially unchanged even if the conductivity in the passive regions changes. Hence an 'expulsion' of a magnetic field depends essentially on the conditions in the active regions.

Applying this to our problem, we conclude that the 'expulsion' of an externally energized magnetic field in a cloud is a complicated process which can be understood only if we know the conditions in the whole circuit. The circuit may keep the cloud magnetized until the power of the whole circuit is dissipated there. Should, however, gravitational forces begin to dominate the contraction of the cloud, a conversion of gravitational energy into circuit energy (and hence magnetic energy) is likely to take place.

The cloud will thus remain magnetized. Hence, in order to prove theoretically that the magnetic field is 'expelled', it is necessary to:

(a) map the whole circuit in which it flows. This may be difficult because the circuit may very well include a large part of our galaxy,

(b) prove that enough energy to keep the current alive is *not* supplied to the circuit.

Hence it is illegitimate to make the common categorical statements that electromagnetic effects can be neglected in the evolution of an interstellar cloud.

An analogy with the Earth is often made: it is stated that the auroral currents flow only in the upper atmosphere, and hence hydromagnetic effects are not essential in the denser regions. This conclusion is misleading. First of all, the auroral currents *do* produce magnetic fields in the whole atmosphere (and in the solid Earth), but the auroral currents are not strong enough to make this field very important. Further, because the Earth's atmosphere is in an essentially static state, even hurricanes do not have so much energy that they could produce appreciable hydromagnetic effects. Contrary to this, in a contracting dark cloud, a gravitational energy of the order of electron volts per atom, or in the later phases even orders of magnitude more, is released.

V.4. Electromagnetic Effects Aiding the Formation and Contraction of Clouds

Electromagnetic forces in cosmic clouds have often been supposed to exclusively counteract a gravitational contraction. This conclusion is based on the assumption that a cloud has a fairly homogeneous magnetization. Like so many other homogeneous models, this should be revised. We know that the general effect of electromagnetic forces could just as well aid, or even cause, a contraction (IV.8.3 and IV.8.4). We can very well assume that an intergalactic cloud is formed from a dispersed interstellar medium by the pinch effect (Alfvén and Carlqvist, 1978). Further development of the cloud can be influenced at least as much by electromagnetic forces as by gravitation, even if at the last stage, gravitation dominates. The cloud may contain small active plasma regions in which most of the energy exchange takes place. These regions may convert gravitational and kinetic energy into electromagnetic energy (III.1.3), so that they are self-supporting.

V.5. Chemical Differentiation in the Primeval Cloud

A cosmic medium has a general tendency to *differentiate chemically*. The condensation of dust grains consisting of the least volatile components of the medium, leads to chemical differentiation. Dust grains above a certain size are not appreciably effected by electromagnetic forces, and hence their motion differs from the motion of an ionized diffuse medium. In IV.3 it was shown that there are plasma processes which cause chemical differentiation of space plasmas; the most important are:

(i) electric currents producing regions of different chemical composition with different ionization potentials,

(ii) diffusion processes active in the boundary region between hot and cold plasma regions,

(iii) differentiation process depending on mass, which gives an isotope separation effect.

As we have seen in II.4, there are enhanced contrast photographs of interstellar clouds showing a pronounced inhomogeneous structure with a network of bright regions presumably due to electric currents. These currents produce thin active regions which interlace the extensive, and usually very cold, regions. In principle, this is the same structure of cosmic plasmas that was derived from magnetospheric studies. In V.6 it will be shown that strong galactic currents are likely to flow in interstellar clouds. This means that two of the chemical separation mechanisms discussed, viz., electric current activity and diffusion at the boundaries between hot and cold regions, are likely to be active. Hence it is reasonable to assume that a cloud of this type has a cellular structure, although not necessarily of the same kind as discussed in II.10 and VI.1.3.1.

Assuming that the primeval cloud from which our solar system originated was a dusty dark cloud, we may conclude that when the sun was forming it was likely to be surrounded by a large number of small cloudlets with different chemical compositions. As long as they were ionized they could not fall in towards the sun because of magnetic fields preventing infall. However, dust grains large enough not to be affected appreciably by the prevailing magnetic fields, would fall in (IV.7.2).

Hence, our study of star formation from interstellar clouds leads to the following model of the very early conditions during the formation of the solar system: the Sun

(i) is already formed by accretion of gas (mostly hydrogen) around a small core of heavy material originating from the dust of the cloud. A similar conclusion, but with a different motivation, has been drawn by Prentice (1976).

(ii) is magnetized by the same mechanisms which produce magnetic fields in all celestial bodies (IV.5.2).

(iii) is not necessarily very hot, actually it may be a very large deuterium-burning star, and it is surrounded by a very low-density region because much material has already fallen into the Sun during its formation.

More distant parts of the primeval cloud have not yet fallen into the Sun, because they have been prevented by the magnetic field. However, from these outer parts the rain of dust, out of which the solar core was formed, is continuing uninterruptedly.

When the temperature in the surrounding plasma decreases, the partially ionized cloudlets of which the primeval cloud consists, will be deionized. Hence the magnetic field will not prevent them from falling in any longer. Consequently, there will be an injection of neutral gas into the solar surroundings, and the chemical composition of this will vary, depending on from which cloudlet it comes.

This picture is derived from the observed properties of interstellar clouds with our knowledge of plasmas in the magnetospheres as a general background. Unfortunately this cannot be checked by *in situ* measurements (which should have been made several billion years ago in order to be relevant!). however, as we shall see in V.7, it leads to a *band structure of the solar system similar to the observed band structure*. This is a strong support for the interpretation we have arrived at.

We can also present this argument in a somewhat different form: The model of the early conditions in the solar system which we have derived from a study of the evolution of interstellar clouds is reconcilable with the model which was already derived earlier from the analysis of the present structure of the solar system. In both cases the investigation is based on the new approach to the physics of cosmic plasmas, including the key role of plasma effects, and inhomogeneous models. It is an approach starting from observational facts and it is essentially free from *ad hoc* hypotheses. For details see Alfvén and Carlqvist (1978).

V.6. Intrinsically Produced Currents

In the Earth's interior there is a current system which produces the Earth's general magnetic field. The mechanism by which this occurs is referred to as 'the self-exciting dynamo' (IV.5.2 and IV.5.3).

We also know that Jupiter and Saturn possess similar self-exciting dynamos in their core.

The only requirements for a self-exciting dynamo (of whatever structure) seems to be that the body rotates and that there is enough energy release in the core. These prerequisites seem to be satisfied in protostellar clouds. They certainly rotate, and when the cloud contracts, energies of electron volts per atom or sometimes much more are released.

This is many orders of magnitude greater than can be imagined to be available in the planetary interiors.

Therefore, as planets possess such dynamos, it seems all the more likely that protostellar clouds also should have them.

V.7. Band Structure and the Critical Velocity

Since the time of Laplace, it has been believed that the solar system must have formed with a homogeneous disc at an intermediate state. There has never been any good arguments in favor of this view. In fact if we plot the smeared-out density distribution in the planetary and satellite systems, we find that this varies drastically. Indeed, there are intermediate regions with densities of $< 10^{-6}$ of that in the most dense regions, and even in the latter, there are variations of orders of magnitude (see Figures V.1 and V.2). It is evident that homogeneous models are as misleading in this field as they are in so many others. The *extremely large variations in mass* accumulated at different distances from a central body is an important fact which must be taken into consideration when accounting for the evolution of the solar system.

Since the matter out of which the secondary bodies (satellites or planets) are formed must be derived from matter falling towards the central body, it is natural to plot the secondary bodies with their gravitational potential, against the velocity they attain if falling from infinity. As this was done in Alfvén (1942) it was discovered that the solar system has a pronounced *band structure* which, if not coincidental, must be of basic importance for the understanding of how the solar system was once formed. See Figure V.3.

The band structure of the solar system can be understood if there is a mechanism which stops matter from falling in from large distances when its velocity has reached certain critical values, v_c. From the band structure diagram it was evident that these values for the most common elements corresponded to the relation

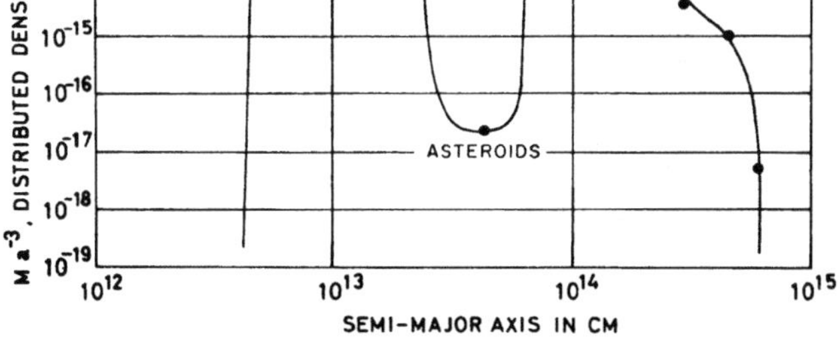

Fig. V.1. Distributed density versus semimajor axis for the planets.

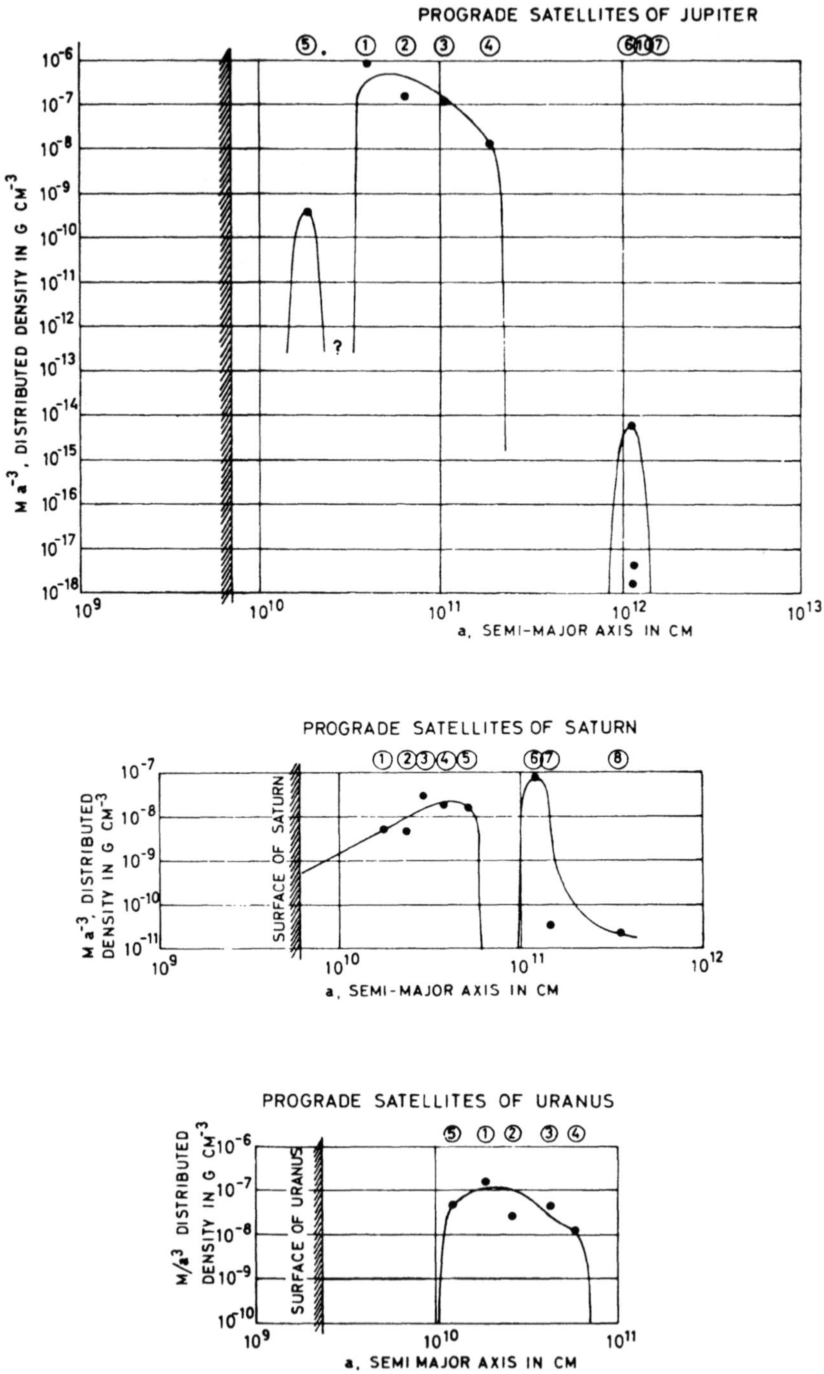

Fig. V.2. Distributed density versus semimajor axis for the prograde satellites of Jupiter, Saturn and Uranus.

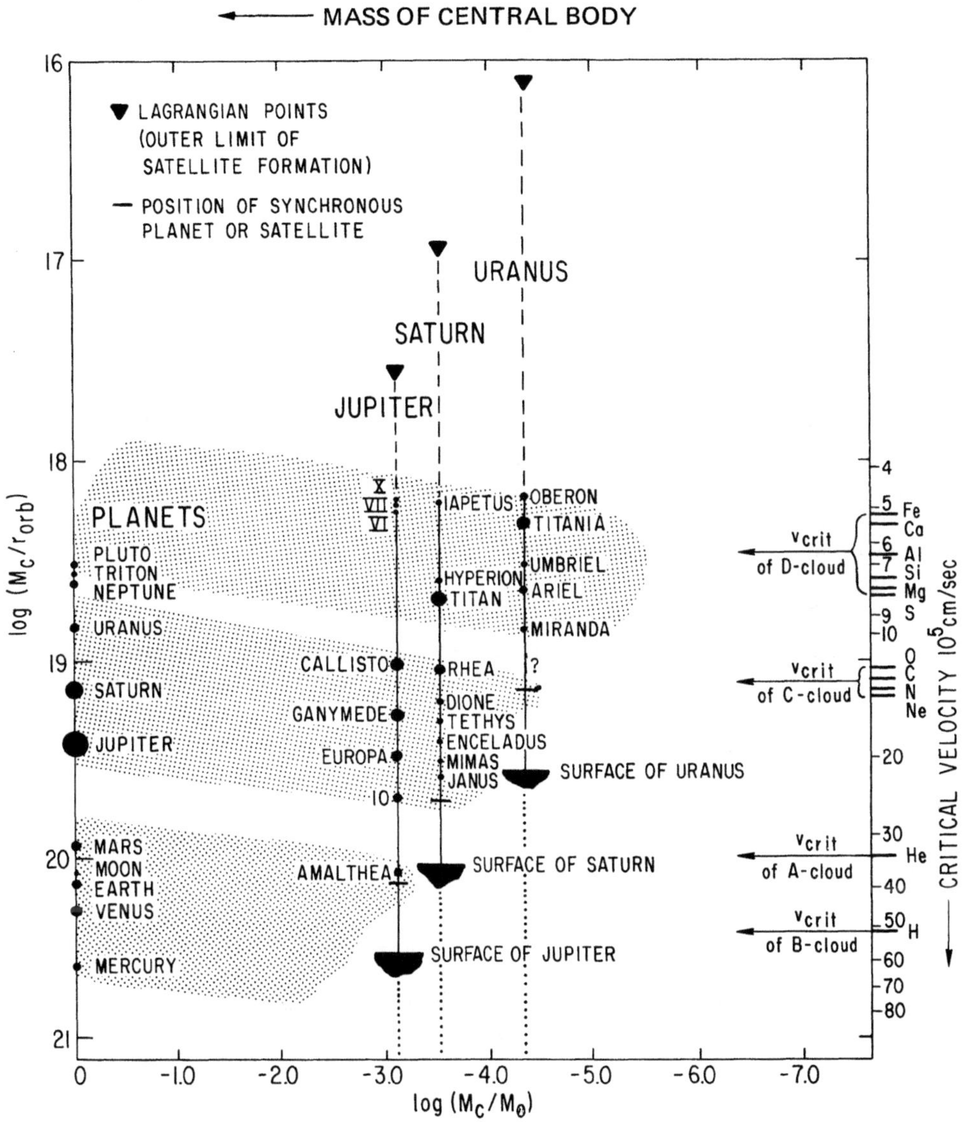

Fig. V.3. Gravitational potential energy (T) as a function of the mass of the central body for the planetary and satellite systems. The right-hand ordinate showing *critical velocity* affords comparison of T values for the planets and satellites with v_{crit} values for abundant elements.

$$\frac{mv_c^2}{2} = eV_{ion} \tag{1}$$

(m mass of atom, V_{ion} ionization potential). A natural explanation would be that when a falling non-ionized gas has obtained the same energy as is necessary for ionization, it will be ionized.

However, when the theory of ionization of a falling gas (in relation to, e.g., a thin plasma resting in the magnetic field of the central body) was developed according to the

the classical theory, it was found that no strong effect should be expected when v_c was reached. As the classical theory could not be considered as sacrosanct, an experimental investigation was started as soon as thermonuclear technology made this possible. The experiments demonstrated that as soon as *the relative velocity between a non-ionized gas and a magnetized plasma exceeded v_c, a very strong interaction is observed.* This was called the 'critical velocity' phenomenon, and has been extensively investigated both experimentally and theoretically (IV.6). *The critical velocity is the only physical phenomenon discovered after being predicted by theories of solar system formation.* Further, based on the band structure, De *predicted the existence of a ring around Uranus* which was discovered several years later (see B. De, 1978). This is the only time an essential new feature of the solar system has been predicted by a theory of solar system formation.

The Jovian ring was not explicitly predicted, but it is located in one of the bands of the band structure diagram which was constructed decades before its discovery.

Further, since the construction of the band structure diagram, the following new celestial bodies have been discovered: the Saturnian satellites Janus and 1979 SJ, the Uranian satellite Miranda, and one Jovian satellite. *All of the regular satellites fall in the bands of the diagram.*

A large number of new asteroids have also been discovered. They do not change the mass–distance diagram of the asteroids. They do not contradict the earlier results, but neither do they add support to the band structure picture.

The band structure was initially achieved from a study of the present structure of the solar system. But it may also be derived from the study of interstellar clouds as has been shown in V.5, based on what we know now of cosmic plasmas in general. This knowledge is essentially gained from *in situ* measurements in the magnetosphere, but also from laboratory studies. We therefore conclude, that the band structure cannot be due to coincidence. *Its reality and significance are very well confirmed.*

The band structure is one of several phenomena which demonstrate the importance of plasma phenomena at the early formative phases of the solar system.

V.8. Solar System in Formation

In this section we sketch the evolution of a solar system in formation, and thereafter we discuss what kind of observations could reveal such a process.

V.8.1. INTERSTELLAR CLOUDS

From the general properties of space media, which the last few years of magnetospheric investigations have clarified, we can draw the following conclusions about the properties of interstellar clouds.

(1) As shown in IV.8 interstellar clouds may very well *be formed by electromagnetic forces* producing a contraction ('pinch effect') in a region in which initially the density is very low. When the cloud is formed, its density may be increased by the same effect up to the limit at which a Jeans collapse sets in, either in the whole cloud or in a small part of it. Hence there is no need for a 'triggering mechanism' (like shock waves, etc.) to produce a gravitational collapse. This does not rule out that in certain cases triggering mechanisms may be important, but they have to be treated according to the new formalism.

(2) *Chemical differentiation* should be a quite normal process in cosmic clouds and may be essential for the chemical differentiation observed in the solar system (V.5). Certainly, a chemical differentiation would be counteracted by turbulence. However, as we have mentioned, there is no certain indication that turbulent mixing is of any importance in the magnetospheres (IV.4). Wildly varying values of plasma parameters and broadening of spectral lines may be due to waves or to filamentary structures. There seems to be no certain indication that large scale turbulence is of major importance anywhere in cosmic plasmas (IV.4).

(3) *Filamentary structure.* There is both theoretical and observational evidence for the view that filamentary structure is a characteristic feature of cosmic clouds (II.4). And since current–carrying filaments and surfaces play such an important role in the magnetospheres, we should expect similar phenomena in cosmic clouds. What has been said eliminates any argument that the evolution of a cosmic cloud can be treated with neglect of electromagnetic effects. This seems to be in agreement with what many authors have found in different ways (Larson, 1978).

(4) As currents produce *regions with much higher densities than the average* (II.4.4), the condensation into dust grains and their increase in size may proceed orders of magnitudes more rapidly than is indicated by calculations based on average plasma densities.

V.8.2. FORMATION OF PROTOSTARS

The generally accepted view is that protostars are formed by gravitational collapse. This is probably true to some extent. However, there are at least *two effects* which may assist the collapse so that also very small clouds (one solar mass or less) can collapse. These make trigger actions (by shock waves etc) and fragmentation in the collapsing cloud unnecessary (but it does not rule out such mechanisms).

(1) The first of these is *the electromagnetic contraction* we have discussed in V.4. This may increase the density of a cloud – even a small cloud – and bring it up to the limit of gravitational collapse.

(2) The other mechanism should take place in a *cloud consisting of a dusty plasma* (Horedt, 1976). If the dust at least partially consists of particles which are so large that their motion is not determined by electromagnetic effects, these will accrete to a dust ball, which serves as a core around which the gaseous components of the cloud can collpase. (For details see Horedt (1976) or Alfvén and Carlqvist (1978).)

Both these mechanisms do not necessarily produce an immediate collapse of the whole cloud. If the current density in the cloud is inhomogeneous – as is normal in cosmic plasmas – it may cause a contraction of a number of cloudlets until they reach the Jeans limit. Hence the cloud will be fragmented into a number of '*stellesimals*' which later form a protostar by essentially the mechanism by which planets accrete from planetesimals. Similarly, if a dusty cloud is inhomogeneous (as most clouds are likely to be) so that the gravitational potential in it has a number of local maxima, a number of cores for local condensation are produced. Also in this way stellesimals may be formed which later accrete to a protostar. Hence the fragmentation of a cloud may take place *before* the contraction, not necessarily *during* it. This mechanism which has been studied in detail by Horedt, leads to the same mechanism for the formation of stars, planets and satellites. (It should be observed that one of the difficulties which Horedt is concerned about, viz. the low density of the interstellar clouds, is not very serious if the current produced inhomogeneity is taken into account.)

V.8.3. PLANET-SATELLITE FORMATION

If a protostar is formed in this way it seems likely from a theoretical point of view that it is surrounded by an almost void region which separates it from the rest of the 'primeval cloud' out of which the protostar has formed. As mentioned earlier electromagnetic effects at this stage are decisive. There are two general approaches for explaining the planet-satellite formation, the hydromagnetic model and the Laplacian.

(A) The *hydromagnetic model*. If we accept that a part of the 'primeval cloud' from which the protostar is formed, is 'hung-up' – perhaps by the electromagnetic effects outside the void region around the protostar – and the matter of this cloud rains down towards the protostar during a long time, the planets may be formed by a semi-stationary process during a long period. This situation is analysed in detail in ESS. Basically the same process takes place around the planets leading to the formation of satellites. The main steps in the process are:

(1) Chemically differentiated small clouds of neutral gas and dust fall towards the central body through a region where the gas (plasma) density is low (typically coronal densities). When it reaches the *critical velocity* the gas is ionized and stopped. This explains *the band structure of the solar system*.

(2) There is a *transfer of angular momentum* from the central body to the plasma by a current system of the type which is well-known from magnetospheric studies (III.3). This transfer decelerates the angular velocity of the central body and gives the plasma the angular momentum which is later possessed as orbital momentum of the secondary bodies.

(3) There is a condensation from the partially corotating plasma, and from infalling grains trapped by it, which results in *planetesimals*. These accrete to secondary bodies. Their masses grow slowly during the whole period during which matter falls in from the primeval cloud. When the mass of the primeval cloud is exhausted – or dissipated – the growth of the secondary bodies stops. They have reached almost their present state. This process is illustrated in Figure V.4.

This sequence of processes leads to the formation of the planets orbiting around the Sun and is later repeated in miniature around the planets where it leads to the formation of (regular) satellites.

(B) The other approach is the *Laplacian model* which is a non-hydromagnetic process of solar system formation. This is divided into two branches: the Kuiper–Cameron collapse models (Cameron, 1978) and the planetesimal model. For reasons summarized by Larson (1978) there are serious objections to the collapse model (see also ESS).

In the Laplacian type theories – including their planetesimal versions – at least all the present mass of the planets is located in a nebula in the solar system region. One of the consequences of this is that the density of the dispersed medium (before the formation of planetesimals) necessarily is very large, so that the mean free path must be small and the evolution very rapid. The result is a flat disc, e.g., of the Goldreich–Ward type (Goldreich and Ward, 1973). If such models claim to be of permanent value it is unavoidable that they are developed into general theories of the formation of secondary bodies. This means that they have to account for the formation of asteroids. It will necessarily be difficult to

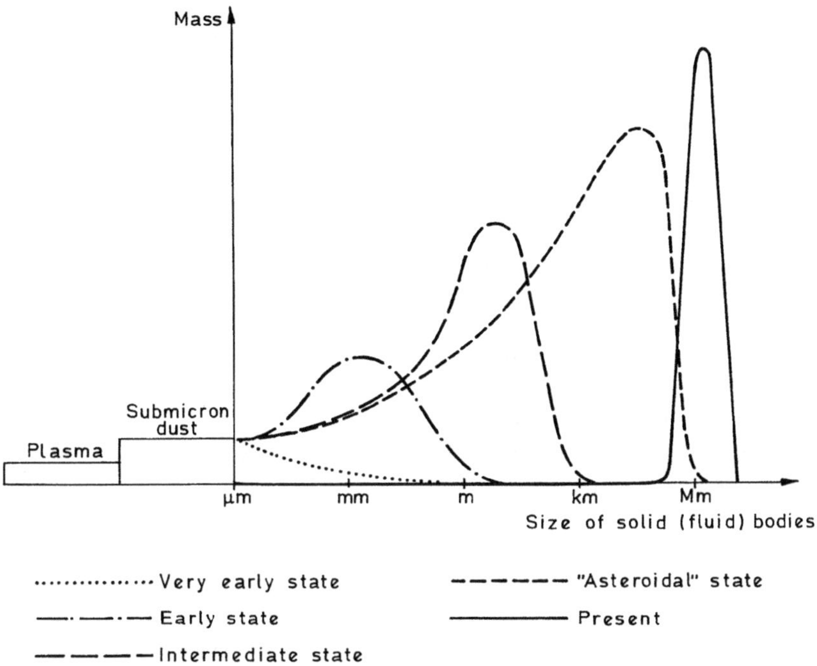

Fig. V.4. Evolution of planetesimal accretion. Infalling grains are trapped by the corotating plasma and accumulate to planetesimals. These accrete to secondary bodies which grow, and form planets or asteroids.

explain their large eccentricities and inclinations. (This presents no difficutly in the hydromagnetic case, because the asteroidal state is an intermediate state in the formation of planets and satellites.)

V.8.4. OBSERVATION OF SOLAR SYSTEMS IN FORMATION

Attempts have been made to identify observed astronomical objects with solar systems in formation. If these are successful we will obtain important information to help the theoretical understanding of the formative processes. In particular it will help us to discriminate between the non-hydromagnetic (Laplacian) approach and the hydromagnetic models.

From an observational point of view the Laplacian (in its planetesimal version) and the hydromagnetic approaches should probably look rather similar in the late phase. In both cases we have a large number of dust grains in essentially Keplerian orbits around a star which should be in a pre-main sequence state and to some extent obscured by dust.

In the early phases of evolution, the characteristics should be very different. The Laplacian idea is well-known from many papers and so are the attempts to verify the Laplacian model with observations. The hydromagnetic approach has so far not received a comparable interest. The latter should be characterized by:

(i) A star of a pre-main sequence type surrounded by a 'primeval cloud' (of the size 10^{13} m or more) consisting of matter in an ionized state, together with dust, a state similar to a more or less dusty H II region. The plasma nearest the star is in a 'chromospheric' state: highly inhomogeneous and penetrated by prominence-like filaments, with rapidly varying currents. 'Chromospheric' spectra should be emitted.

(ii) Dust discs consisting of planetesimals in Kepler orbits with semi-major axes between 10^{10} and 10^{13} m (if the star is of solar mass), and these are coexisting with the plasma.

(iii) At the same time there is a rapid infall of matter. This is non-ionized until it has reached the critical velocity (52 km s^{-1} for hydrogen), but after ionization it will emit the same kind of spectrum as H II-regions do. The infalling velocity should be in the range of 50 and several 100 km s^{-1} which should give a corresponding Doppler shift. The maximum velocity is given by the escape velocity of the star.

(iv) During the planet formation epoch, and probably mainly before there is also an emission of solar wind type, with comparable velocities. Most of this radiation will originate in the polar region of the stars.

T Tauri stars seem to be possible candidates for solar systems in formation, following the hydromagnetic description.

V.9. Hetegony and the 'Hetegonic Principle'

The origin and evolution of the solar system constitutes a well-defined and important field of research which still does not possess a generally accepted name. As the basic problem is the process by which celestial bodies (like the sun and later the planets) formed *companions* (like planets and satellites) around them, the term 'hetegony' (from Greek *εταιροσ* – companion) has been suggested (ESS, 1.2). If we accept that dark interstellar clouds are the source of protostars, this field of research should be considered as a branch of hetegony.

There are a large number of papers claiming to explain the origin of the terrestrial planets, but disregard the fact that neither the giant planets nor the satellites can be explained by a similar mechanism. There are others claiming to explain special features of a satellite system without mentioning that there are similar features in the planetary system which obviously cannot be explained in the same way.

As has been demonstrated in a number of papers (Alfvén, 1942, 1943, 1946, 1954, 1963: Alfvén and Wilcox, 1962; and especially ESS), the structure of the planetary system and the satellite systems are so similar in essential respects that we should aim at explaining them by a common theory (the 'hetegonic principle'). This principle actually derives from Galileo, who, after discovering the Jovian satellites in 1610, proclaimed that they were a miniature planetary system.

CHAPTER VI

COSMOLOGY

VI.1. The State of Cosmology

VI.1.1. HISTORICAL SURVEY

In the preceding chapters we have seen how laboratory research and *in situ* measurements in the magnetospheres have changed our views of cosmic plasmas in a rather drastic way. We shall here analyse what consequences – if any – this has for the cosmological discussion. As we shall see the issues are whether the universe is homogeneous or inhomogeneous, and finite or infinite. The controversy is in certain respects similar to the controversy 400 years ago although the present universe is about 10^{15} times larger.

Cosmology has always been – and will by definition always remain – a borderland between science and philosophy – some would say religion. During the epoque of the Ptolemaian cosmology the region of the crystalline spheres divided the world into the mundane world 'sub luna' – of a size a few times the size of the Earth – and the divine world beyond the sphere of the fixed stars. Science, for example in its Aristotelean version, explained the mundane world – at least to some extent – but not the heavens.

The Copernican–Galilean revolution smashed the beautiful crystalline spheres, and Newton demonstrated that the same laws of mechanics that ruled in the laboratory – and in his garden – also applied to the 'celestial mechanics' which govern the motions in the solar system.

This lead to a cosmological view of an infinite universe ruled by Newton's 'universal' law. The framework was of course Euclidean. The universe was supposed to have a large scale homogeneous structure, in spite of the fact that the regions which could be studied with some degree of certainty were strongly inhomogeneous.

There was no compelling reason to assume a large scale homogeneity, but as homogeneous models are the simplest possible scientists usually have a tendency to assume homogeneity until there is clear evidence for inhomogeneity.

The Olbers paradox demonstrated that a homogeneous infinite Euclidean universe was not acceptable. This difficulty could be solved by dropping either the homogeneity or the Euclidean framework. Charlier chose the first alternative and postulated that the universe had hierarchical structure. He showed that if the average density ρ obeyed the law

$$\rho = kR^{\alpha} \quad \text{with}\ \alpha < -2, \tag{1}$$

where k is a constant and R is the linear size of a structure, the Olbers paradox was solved.

The other alternative was made possible by the Minkowski geometry on which Einstein based his general theory of relativity. This allows the universe to have a finite volume. The Friedmann model which is one of many relativistic models was used by Lemaitre as a basis for a cosmology which also solved the Olbers paradox. Moreover it claimed to explain the Hubble expansion. The great advocate for this model (or rather a slight modification of it) was Gamov, who claimed that it could explain all the main properties of the universe: besides the Hubble expansion also the synthesis of all elements, a cosmic background radiation, etc. Largely because of Gamov's dynamic personality this 'big bang cosmology' has now become the generally accepted cosmology.

The Hubble expansion is measured by the redshift of galaxies. *Assuming* that the redshift is exclusively caused by a *longitudinal* velocity v of a galaxy at the distance R, we can define the *Hubble parameter*

$$H_0 = v/R\,\mathrm{s}^{-1}. \tag{2}$$

Usually the Hubble parameter is given in km Mpc^{-1} s^{-1} so that a value of for example $H_0 = 100\,\mathrm{km\,Mpc^{-1}\,s^{-1}}$ means $H_0 = (10^{10}\ \mathrm{years})^{-1} = 3.3 \times 10^{-18}\,\mathrm{s}^{-1}$.

As according to the big bang theory the Hubble parameter *should* be constant (in space), it is usually called the *Hubble constant*, a biassed term which will not be used here. Further, for the region of space which takes part in the Hubble expansion we should use the old term *metagalaxy*. According to the big bang hypothesis this is the whole *universe*, a term which its advocates use as a synonym for metagalaxy. The *Hubble time*

$$t_H = H_0^{-1} \tag{3}$$

is usually referred to as the 'age of the universe' and the *Hubble distance*

$$R_H = cH_0^{-1} \tag{4}$$

the characteristic length.

Friedmann's model is basically a four-dimensional homogeneous model which means that it belongs to another category than the three-dimensional inhomogeneous models we have discussed in the preceding chapters.

This does not mean that there necessarily is a sharp conflict between these views. The Friedmann model can very well accomodate inhomogeneous regions as long as they are 'local' and do not invalidate the large scale homogeneity. On the other hand, observations very far out are not yet accurate enough to decide whether inhomogeneous models are valid even in those very distant regions which the Friedmann model claims to describe. Hence our aim is, at least to some extent, to find where the boundary is between 'local' and 'large scale' phenomena. In a way this is analogous with the old boundary between the mundane phenomena and the celestial phenomena, which were believed to be so different that basically different laws were governing them.

In this chapter we shall examine to what distances the inhomogeneous Euclidean description is preferable to the big bang description. If it is found that a Euclidean description is valid all the way out to the Hubble distance R_H, this means that the observable universe *is much larger* than the big bang hypothesis claims. Whether still further out the universe is closed and obeys a general relativistic model, is left as an open question (see VI.6).

Lemaitre was not only a very competent scientist with general relativity as his speciality, but also a prominent member of the Catholic hierarchy. To him personally his theory of 'l'Atome primitif' was an ingeneous synthesis of general relativity and the Thomistic dogma of a creation *ex nihilo* (although he never expressed this explicitly in his papers).

To many people the big bang idea is attractive in the same way, being a synthesis of astrophysics and the dogma of a creation *ex nihilo* (see, e.g., Jastrow (1978)). To some extent it has also conserved a religious intolerance toward those who question whether its foundations are scientifically justifiable.

VI.1.2. THE BIG BANG HYPOTHESIS

In the sixties the cosmological discussion was centered on the question whether the big bang hypothesis or the continuous creation hypothesis was the most promising approach. The discussion demonstrated that there were fatal objections against the latter, but no decisive objections against the former were presented forcefully enough. The result was that effort was concentrated on the development of the big bang model and in wide circles this has now been accepted as the final solution to the cosmological problem.

In recent time there has been remarkably little discussion of whether the basic big bang hypothesis is correct or not. Because of its obvious advantage compared to the continuous creation model it was generally assumed that it must be correct. Authoritative presentations of the modern view of the evolution of the universe, for example, Sciama (1971) and Weinberg (1977), take for granted that no other views are worth considering, as does the majority of the cosmological establishment. Observations are primarily interpreted according to big bang formalism, and the large body of observations which are not in agreement with it are either accounted for by numerous *ad hoc* hypotheses or simply neglected. This applies, for example, to the early claim by Ambartzumian (1958) and later by Arp (1978) that there are ejections from galaxies which demonstrate the existence of 'an unknown force' which counteracts gravitation. The problem of the energy source of quasars and quite a few other objects which release large quanitites of energy, is so important that it must be incorporated in any consistent cosmology but the big bang hypothesis has so far given no answer.

Further, de Vaucouleurs (1970) has demonstrated that observations support a 'hierarchical cosmology' of the Charlier type.

Contrary to the very strong views of most cosmologists it seems highly appropriate to initiate a serious discussion about the validity of the big bang hypothesis. This is especially true in view of the fact that – as we have seen in the preceding chapters – *in situ* measurements of space plasmas have changed our views of cosmic physics, and that this change may include the approach to cosmology.

Criticism of the big bang hypothesis should not be interpreted as an attempt to revive the continuous creation hypothesis. However, the advocates of this hypothesis (especially the Burbidge group) have presented several objections to the big bang hypothesis which deserve much more attention than they have received.

If the big bang theory is shown to be unacceptable, it will leave a 'cosmological vacuum', which necessarily must be filled by another approach. One of the candidates is the Klein model which is based on the Dirac symmetry between matter and antimatter. This approach is of interest only if our metagalaxy is symmetric with regard to matter–antimatter. There is no absolutely convincing proof of this but there are a large number of strong arguments in favor of this view, as will be shown in VI.2.

In this chapter we shall sum up the objections against the big bang hypothesis. It will also be demonstrated that there are a number of arguments in favor of a symmetric approach to cosmology. However, it should be stressed that the criticism of big bang is largely *independent* of the arguments in favor of the symmetrical approach.

VI.1.3. HOMOGENEOUS AND INHOMOGENEOUS MODELS

As we have seen in the preceding chapters, *in situ* measurements of solar system plasmas have demonstrated that their properties are drastically different from what was generally supposed earlier (Fig. I.2). The earlier approach to plasma physics still dominates astrophysics, including cosmology. Obviously, this represents an unstable situation: it is impossible to claim that cosmic plasmas outside the reach of spacecraft should have properties which are basically different from plasmas in the regions which are accessible to high quality diagnostics (cf. Figure I.3). Hence, we can be sure that a revision of the astrophysics of more distant regions, including cosmology, will take place in the near future (Fälthammar *et al.*, 1978).

At present a large number of astrophysical plasma problems are treated with 'homogeneous' models (the properties of a plasma are assumed to be the same in a large volume, or to vary in a continuous way with space coordinates and time). The motivation for such models is to some extent that such assumptions are necessary in order to make the problem tractable from a mathematical point of view. However, this is not a valid argument because – as Fresnel (1826) put it "Nature does not care for analytical difficulties."

VI.1.3.1. *Cellular Structure of Space*

From the cosmological point of view, the most important new space research discovery is probably the *cellular structure of space*. As has been seen, in every region of space which is accessible to *in situ* measurements, there are a number of 'cell walls', sheets of electric currents, which divide space into compartments with different magnetization, temperature, density, etc. In Chapter II.10 it was argued that even outside the present reach of spacecraft, space must have a similar cellular structure. If so, half of the different objects in space may consist of koino-matter and half antimatter (II.10 and IV.9). The cosmological consequences of this will be discussed in VI.2.

VI.1.3.2. *Mass Distribution in the Universe*

Hence, there are good reasons to believe that space plasmas are strongly inhomogeneous. If we turn our attention from the plasma regions to the condensed celestial bodies, we also find a drastic lack of homogeneity.

The distribution of stars in space cannot be described by a homogeneous model. A look at the night sky, indicates a high degree of non-homogeneity. To the naked eye, the Andromeda galaxy looks like a diffuse nebula, but photographs of it show that it has the complicated structure of a spiral galaxy. Only as long as it could not be observed through a telescope was a homogeneous model of it reasonable.

The same applies also further out. The real distribution of mass does not seem to be homogeneous. It is in better agreement with de Vaucouleurs' 'hierarchical' model, which is a strongly inhomogeneous model with galaxies forming groups or clusters which are parts of superclusters. The mean densities ρ of such formations (with radius R) obey the empirical law (see Figure VI.1) (de Vaucouleurs, 1970)

$$\rho_V = 6.0 \times 10^{14} R^{-1.7} \,\mathrm{kg\,m^{-3}} \tag{5}$$

valid out to at least 10^{24} m $\approx 0.01 R_H$.

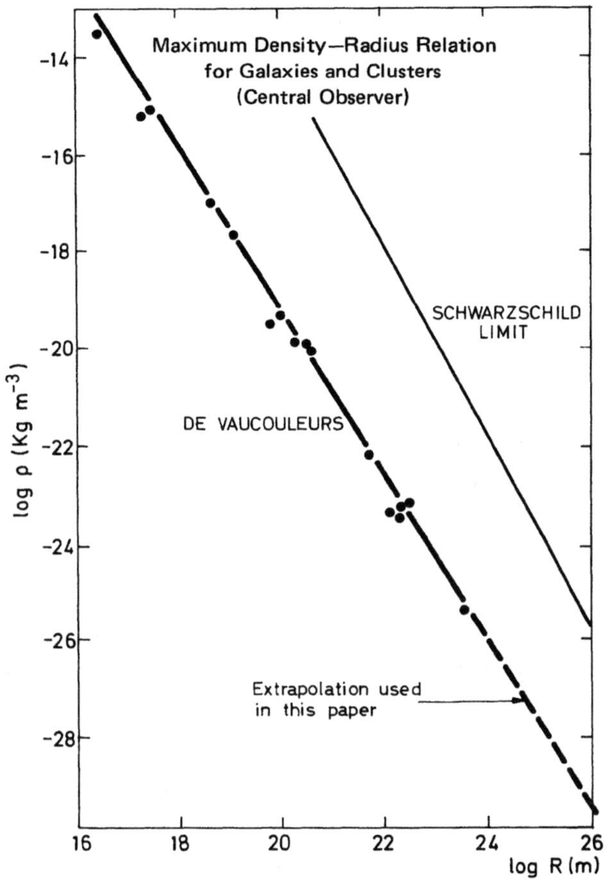

Fig. VI.1. Correlation between average density and radius of spheres centered on galaxies and system of galaxies (de Vaucouleurs 1970). This relation is extrapolated to the Hubble radius. As a comparison the Schwarzschild limit is given. As the observational curve is far below the Schwarzschild curve, it is legitimate to use Euclidean geometry when treating not only galaxies and clusters of galaxies but also the metagalaxy.

When de Vaucouleurs suggested this model, it was objected that the statistical treatment of galaxy counts ruled out any appreciable deviation from homogeneity (Sandage *et al.*, 1972). Investigations by new methods (two-point, and also three-point, correlation) have, however, fully confirmed the hierarchical structure. A survey is given by Groth *et al.* (1977). They find the exponent in Equation (5) to be -1.75. In other papers, Groth and Peebles (1977) and Fry and Peebles (1978) give additional support to the continued clustering hierarchical distribution of de Vaucouleurs. They give the value -1.77 for the exponent. These measurements do not give any certain information about regions outside about $0.01\,R_H$ and do neither disprove nor prove homogeneity further out. However, recent results by Tarenghi *et al.* (1980) have clearly demonstrated the existence of a hierarchical structure further out. This may apply as far as distances corresponding to a redshift* $z = 0.05$.

Radio astronomy observations are also quoted to support the homogeneity of the metagalaxy at large distances. Most of their results in this respect are based on the *assumption* that all the QSOs are at* 'cosmological distances'. As this assumption is questionable (see VI.3.7 and VI.4), also this support for the big bang is eroded.

de Vaucouleurs' law is empirical and there does not exist any theoretical derivation of it. Theoretical laws with which it should be compared is the Charlier criterion $\alpha < -2$ in Equation (1) for the convergence of large scale structures in a Euclidean framework, and the Laplace–Schwarzschild limit

$$\rho_{\text{sch}} = \frac{3c^2}{8\pi G} R^{-2} = 1.6 \times 10^{26} R^{-2}. \tag{6}$$

The value -1.7 or -1.77 for the exponent in the de Vaucouleurs formula is somewhat larger than -2 in the above formula. This means that going outwards we approach the Schwarzschild limit but very slowly. In fact, by increasing the size of a structure by a factor of 10^6 (e.g., from the size of a galaxy to the size of the metagalaxy) we come closer only by a factor of $10^{6(2-1.7)} = 63$ or $10^{6(2-1.77)} = 24$.

In summary, the observational support for a homogeneous density is not convincing. There seems to be rather strong indications of a hierarchical density distribution as far out as reliable observations are available. If we extrapolate the de Vaucouleurs law by one order of magnitude (or perhaps two), we may conclude that the metagalaxy is a member of a Charlier hierarchy, indeed a 'super-supercluster' (Figure VI.1).

VI.1.3.3. *Mass of the Metagalaxy and the Schwarzschild Limit*

Following Gott *et al.* (1974) we put

$$\Omega = \frac{8\pi G}{3H_0^2} \rho_M \tag{7}$$

where H_0 is the Hubble parameter, G the gravitational constant, and Ω is the ratio between the mean density ρ_M and the critical density for closure. If $\Omega \geqslant 1$, it would be difficult to avoid the big bang hypothesis. Its advocates have spent a great deal of energy on finding the 'missing mass', but so far they have had no success. Gott *et al.* (1974) have made a critical analysis of the most reasonable value. They find as an *upper limit* $\Omega = 0.06$. As they assume a homogeneous density, the value should be much smaller with the de Vaucouleurs' mass distribution. In fact, the value of Ω which results from the extrapolation of de Vaucouleurs' formula (5) is of the order of $< 10^{-3}$ (and still smaller with the exponent -1.77). With this value the metagalaxy, the size of which was smaller at earlier epochs (as inferred from the Hubble expansion), could have been as small as 10^{-3} of its present size without being inside of the Schwarzschild limit.

* The relation between redshift $z = \Delta\lambda/\lambda$ and $\beta = v/c$ is $1 + z = (1 + \beta)^{1/2}(1 - \beta)^{-1/2}$.

VI.1.4. THE HUBBLE EXPANSION

VI.1.4.1. *The Hubble Parameter*

As no acceptable alternative seems to exist, we interpret the observed galactic redshifts as due to Doppler effect (longitudinal and transverse). If we neglect the transverse Doppler shift – which is not obvious that we should unless we accept the big bang hypothesis – this means that at present the metagalaxy is in a state of general expansion. For a long time the best measurements of the Hubble parameter seemed to indicate that it is a constant (in space). The belief in its constancy was so strong that it was baptized 'the Hubble constant'. Doubts about this constancy have accumulated, and must now be taken seriously. For example, de Vaucouleurs and Bollinger (1979) find variations between $2.5 \times 10^{-18}\ \mathrm{s}^{-1}$ (70 km s^{-1} Mpc^{-1}) and $3.5 \times 10^{-18}\ \mathrm{s}^{-1}$ (110 km s^{-1} Mpc^{-1})* for galaxies in different directions, whereas the Sandage observations indicate $1.8 \times 10^{-18}\ \mathrm{s}^{-1}$ (55 km s^{-1} Mpc^{-1}). For references see, e.g., Kristian *et al.* (1978) or Sandage *et al.* (1979).

A comparison between these papers, both written by prominent observers, gives a good view of how uncertain the observational results are from which far-reaching cosmological conclusions are drawn.

If the Hubble parameter is not a constant, this means that a linear extrapolation backward in time does not indicate that at any time the matter in the metagalaxy (= 'universe') was concentrated in a very small volume. Instead, a linear extrapolation leads to a minimum size of the metagalaxy which may be quite large. This is shown in Figure VI.2(a), constructed by Bonnevier (1978) from data given by Kristian *et al.* (1978) (in their Table 5). The receding velocities v given by the redshifts of 16 galactic clusters measured and the distances R_n they have estimated are used for reconstruction of the motion in the past of each individual galaxy, according to the linear formula

$$R = R_n - vt, \tag{8}$$

where t is the time reckoned backwards from now. The linear extrapolation is legitimate only if the paths are fairly unperturbed by gravitational effects. This seems probable. In Figure VI.2(a) the value of $|R|$ is plotted as a function of time.

The diagram shows that the clusters were at a minimum distance from each other at about $-t_H$, but this minimum was as large as about 13% of their present distance. Other groups of galaxies or clusters give similar results.

Of course there is a considerable uncertainty in the distance measurements, which means that the discrepancy may very well be due to observational errors. Hence the diagram *does not* disprove the big bang hypothesis, according to which all the paths should converge in a single point at $-t_H$. But neither does the diagram give decisive support *for* it. A Hubble diagram of the same clusters is shown in Figure VI.2(a).

If these results are typical for all measurements of the distance-redshift relation – as they seem to be as judged from a number of similar analyses – the only conclusion we can draw from the observations is that at about $-t_H$ our metagalaxy was smaller than about 10% of the present size. If we accept the large variations reported by de Vaucouleurs and others, the minimum size should have been much larger.

* When American laymen are planning to leave 'feet' and 'ounces' it is perhaps time for cosmologists to leave 'km s^{-1} Mpc^{-1}' as a common unit.

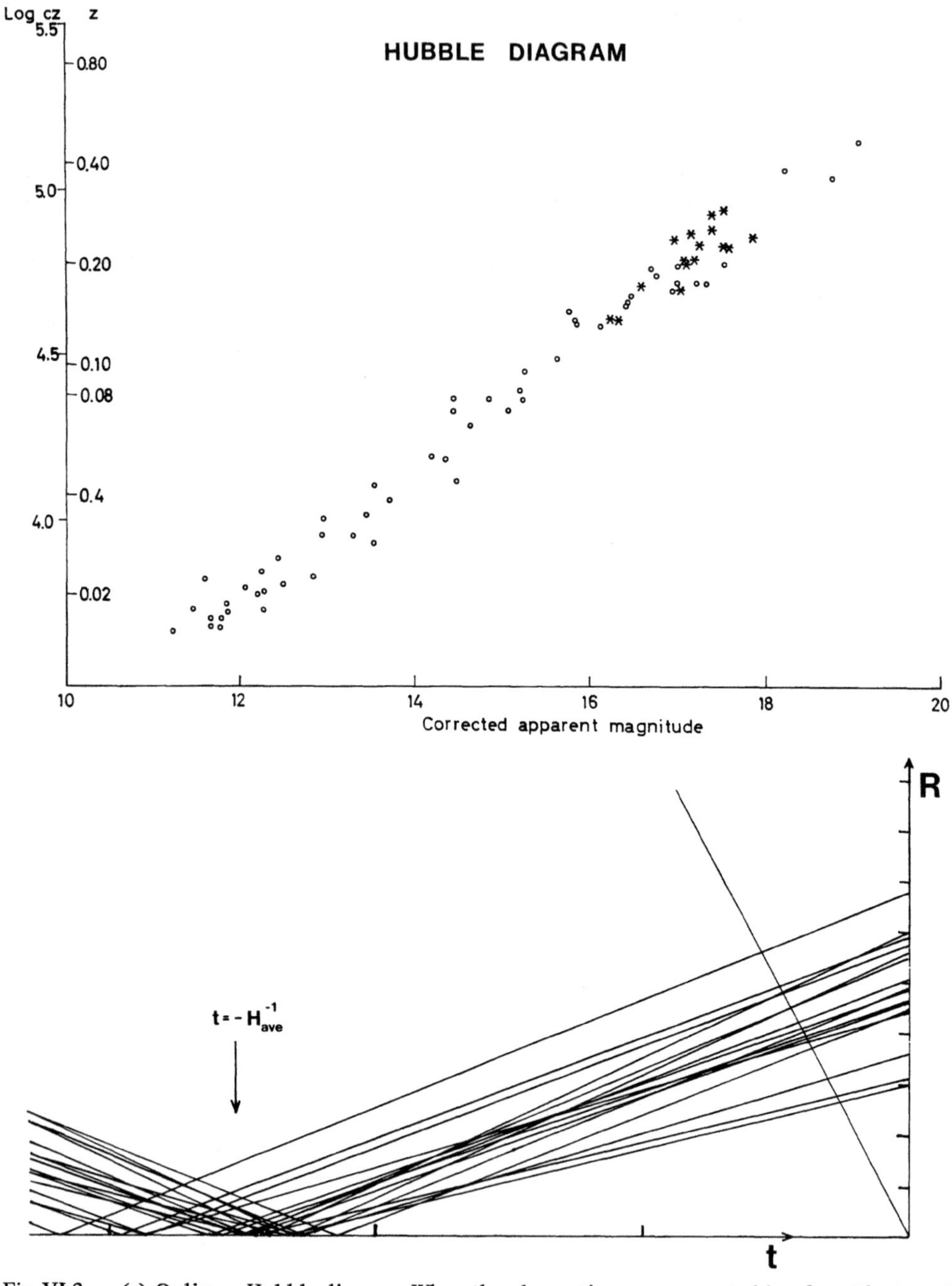

Fig. VI.2. (a) Ordinary Hubble diagram. When the observations are presented in a *logarithmic* scale they appear to show that H_0 is a constant (within the limits of error) and to confirm the big bang hypothesis. (b) Unbiassed treatment in a *linear* scale diagram. The same observational material gives another impression if from each observation the receding velocity V and the observed distance R_n are used to make a linear extrapolation of the distance $R = R_n + vt$, and the value $|R|$ is plotted (Bonnevier, 1978). From the observations it is impossible to conclude that the metagalaxy was ever smaller than about 10% of its present size.

This means that until it has been clearly demonstrated that the Hubble parameter really is a constant, *the big bang hypothesis is not supported by the Hubble expansion.* Certainly, it is always formally possible to introduce *ad hoc* assumptions to explain *any* disagreement between theory and reality. However, as the credibility of the Ptolemaic cosmology diminished with the number of epicycles that had to be introduced, the credibility of big bang decreases with every *ad hoc* assumption which is needed.

VI.1.4.2. *Euclidean Model of the Evolution of the Metagalaxy*

If Ω where $\geqslant 1$, a relativistic treatment of the present metagalaxy would be necessary and the big bang theory probably unavoidable. If we accept the value $\Omega < 0.06$ by Gott *et al.* (1974) it is so far from being closed that a Euclidean description of the present state is a reasonable first approximation. A hierarchical structure, leading to $\Omega < 10^{-3}$ (see VI.1.3.3) means that there is no reason to use a general relativistic treatment except as a small correction. If we put the size and mass of a core of a galaxy equal to 10^{20} m and 10^{40} kg, we are 10^{-6} away from the Schwarzschild limit. If in de Vaucouleurs' formula (5) the exponent were -2 instead of -1.7 (or -1.77), the metagalaxy would not be any closer to the Schwarzschild limit than our galaxy. The smaller absolute value of the exponent brings the metagalaxy closer but only by one or two orders of magnitude (which is still very far from the limit). Indeed, in theories of galaxies it is not necessary to use general relativity, and this holds also for galactic clusters and superclusters. *From an observational point of view there is no need to change from a Euclidean to a relativistic treatment when we go from superclusters to the 'super-supercluster', the metagalaxy.*

Our ambition in this chapter should be to try to reconstruct a state which we will call the *proto-metagalaxy* which is the state which our metagalaxy had at the time of the order of $-t_H$, when its size was about $0.1\,R_H$. See further VI.5.

With this size the proto-metagalaxy was 10 times closer to the Schwarzschild limit than now. But even $\Omega < 0.06$ means that we still are outside the Schwarzschild limit. A Euclidean treatment would be allowed as a first approximation, but a large relativistic correction would be necessary. With $\Omega \leqslant 10^{-3}$, as we have found in VI.1.3.3, we are so far from the limit that the general relativistic correction is only of the order of 1%.

Hence the evolution of the metagalaxy from the state of the proto-metagalaxy to the present state can be described in a Euclidean framework; the general relativity correction can be neglected in a first approximation. (Of course the special theory of relativity is necessary.)

If not under pressure from the big bang establishment, no observer would get the idea of changing the general treatment of the observations when proceeding from 10^{20} m to 10^{26} m.

A treatment of the observational material in a Euclidean framework will give a picture which is clearer and more independent of *any* theory, than a four-dimensional general relativity picture. The cosmological discussion should then start with the integrated observational picture as a launching pad.

VI.2. Coexistence of Matter and Antimatter

VI.2.1. MATTER-ANTIMATTER SYMMETRY

The discovery of the positron and the antiproton has led to the recognition that, in principle, there could be antimatter in the universe. As celestial bodies consisting of

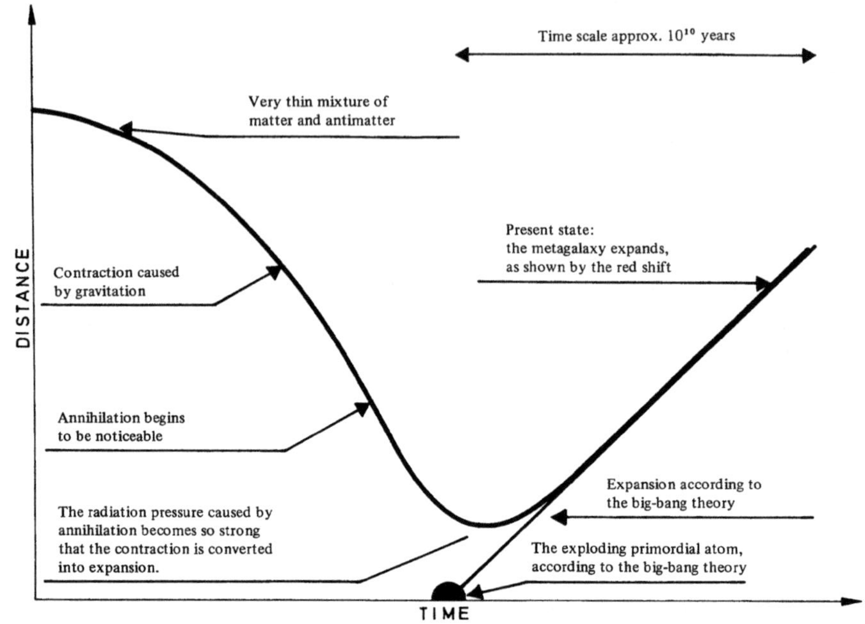

Fig. VI.3. Evolution of the metagalaxy in Klein's model. Gravitational attraction of the original matter–antimatter mixture leads to a condition where annihilation causes the expansion now observed by the redshift.

antimatter emit the same spectra as bodies of ordinary matter (koino-matter) it is impossible to tell from a distance what kind of matter they consist of (see IV.9). The proof for or against the existence of antimatter has to be indirect (Alfvén, 1965, 1971).

Oskar Klein (Alfvén and Klein, 1963) has suggested a cosmology often referred to as the *symmetric cosmology*, according to which there should be equal amounts of koino-matter and antimatter in the universe (Figure VI.3). Although such a symmetry appeals to most scientists, there is naturally a reluctance to discuss a theory which must necessarily lead to a revision of quite a few of the present theories in cosmic physics, including cosmology. In fact, the present theoretical framework has been inherited from the time when it was generally believed that all matter must necessarily be of the ordinary kind. Since it is now evident that, in principle, the universe may be symmetric, it is necessary to go through large parts of astrophysics in a systematic way in order to see what the consequences of symmetry are.

VI.2.2. MATTER AND ANTIMATTER CELLS

We know that a cellular structure characterizes those regions of space which are accessible to spacecraft, but as such structures cannot be detected at a distance, we have no certain information about more distant regions. There is no reason to suppose that their existence should be limited to regions of space where spacecraft have thus far penetrated. As we now begin to understand how the cells are formed we can conclude with a high degree of confidence that both *interstellar space and intergalactic space should in general exhibit a similar cellular structure*. However, the size of this structure is difficult to derive theoretically and impossible to observe directly.

It is easily seen how important the discovery of the cellular structure of space is to the discussion of antimatter in the universe. The demand for symmetry is satisfied if the metagalaxy, or even our own galaxy, is divided into a large number of cells, half of which contain koino-matter (ordinary matter) and half antimatter. The discussion in IV.9 showed that cells of different kinds of matter should be separated by Leidenfrost layers, thin layers of discontinuity containing high energy electrons–positrons produced by annihilation of protons and antiprotons (or other nuclei) at the interphase (Figure VI.4). A theory of such layers developed by Lehnert (1977) shows that under cosmic conditions they need only to be 10^8 m (or about 10^{-8} of a light year) thick. The basic reason is that annihilation produces a sink of koino- and antimatter, leading to a plasma pressure gradient which is balanced by the force from electric plasma currents and a magnetic field. This force pushes the two opposite plasma regions away from each other, and the rate of annihilation is substantially reduced upon reaching a quasisteady balance. As is true for the interphase between the magnetosphere and interplanetary space, such a layer should be very difficult to discover unless a spacecraft penetrates it. The annihilation radiation emitted from it is too small to be detectable with present measuring devices. Thus we cannot exclude that matter and antimatter can coexist in a universe which is pervaded by a system of Leidenfrost layers. This would not bring us in contradiction with any observed phenomena.

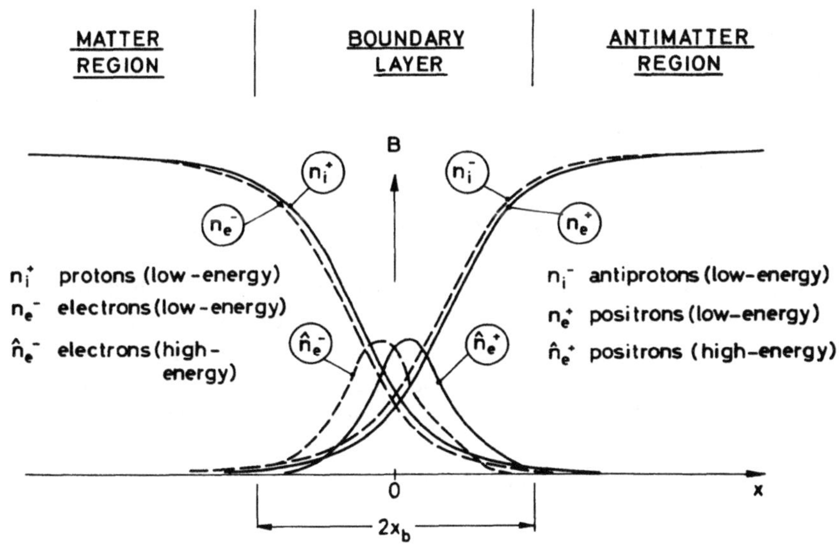

Fig. VI.4. A Leidenfrost layer, separating a region of antimatter from a region of koino-matter (from Lehnert, 1977).

VI.2.3. SIZE OF CELLS: GALACTIC OR SMALLER?

The demand for symmetry can be satisfied in several ways. We may postulate that every second galaxy (or every second group of galaxies) consists of koino-matter and every second of antimatter. If, for example, the Andromeda galaxy consists of antimatter and is separated from our own by a Leidenfrost layer, this would be of considerable cosmological importance but would not affect intragalactic phenomena much. Such a model would on the one hand relieve us from several of the objections to the existence of antimatter, but on the other hand, deprive us from invoking annihilation as a source of energy for a number of observed processes, e.g., the energy release in quasars. As the objections against antimatter are in any case very weak (VI.2.5) and annihilation is urgently needed as a source of energy (VI.3), this is not the most attractive alternative (although it should not be completely forgotten). It seems more interesting to tentatively investigate what the consequences are of every galaxy being koino–antimatter symmetric. This does not require that we specify our model in detail. For example, from this point of view it would be possible that one part of a galaxy consisted of koino-matter and another of antimatter, or that the cells are the size of a cluster of stars. The most extreme case is that the cells have a size equal to the volume occupied by a single star and the system (planets, comets, stellar wind, etc.) belonging to it (say 10^{50} m^3). The most unexpected and shocking result of such an analysis is that *we do not know with certainty whether our closest neighbours in space* (for example, α Centauri) *consists of matter or antimatter.* While leaving other options open we shall, in the following, focus our attention on the alternative of cells of the average size of $\sim 10^{50}$ m^3.

VI.2.4. STRUCTURE OF A SYMMETRIC GALAXY

VI.2.4.1. *Solar system*

With this as a background let us try to make a tentative model of our galaxy. It is easily seen that in our own solar system practically all bodies must consist of koino-matter (Sun, planets, satellites and at least most of the comets and meteoroids). We can, perhaps, not exclude that a few meteoroids might consist of antimatter, but attempts to prove this have not so far been successful. Concerning the comets of which some 10^{10} or 10^{11} are believed to be located in the Oort 'cometary reservoir' at 10^{14}–10^{15} m from the Sun (Figure VI.5), there is not yet any indication that any one of them consists of antimatter, but we cannot exclude that a few of them do. The solar wind which, of course, consists of koino-matter is known to penetrate as far as spacecraft have reached, but whether it also penetrates the cometary reservoir or part of it is subject to speculation.

Even if we accept that all matter in a sphere with a radius of 10^{15} m around our Sun consists of koino-matter, we must note that this distance is only a few percent of the distance to our closest stars (see Figure VI.5). Hence there is ample room for Leidenfrost layers, separating the Sun-dominated region from those belonging to other stars. We are not in conflict with any observational facts if we claim that one or more of our closest stars consist of antimatter. If we claim that in our galaxy every second star consists of antimatter, there is at present no way of proving or disproving this in a straightforward way. Every star in our galaxy should be surrounded by a region of the same kind of matter as itself. The topology of the separating Leidenfrost layers forms an interesting problem which remains to be investigated.

It may look unpleasant to many, to postulate the existence of these Leidenfrost layers everywhere in the universe, but to those cosmic plasma physicists to whom Cahill's discovery of the thin discontinuity at the magnetopause came as a shock, it would be more unpleasant not to postulate it.

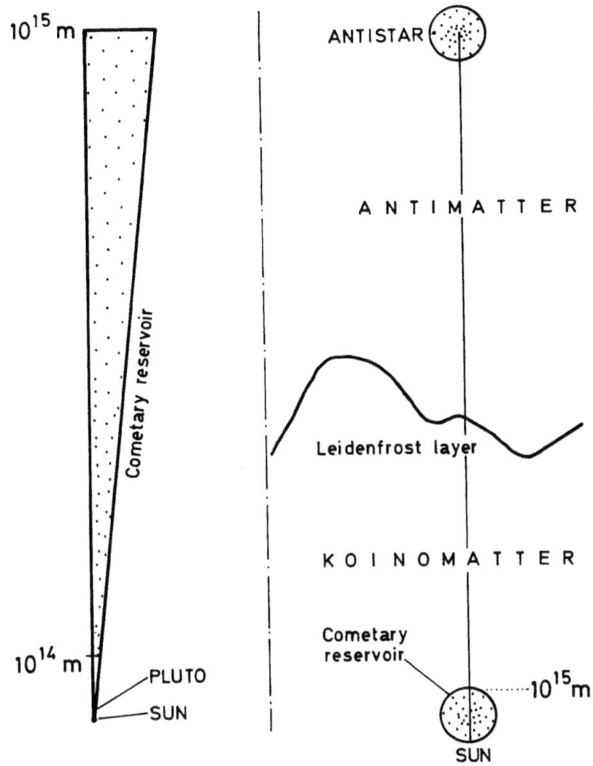

Fig. VI.5. (Left) Cometary reservoir surrounding the planetary system. (Right) Separation between matter around our Sun and antimatter around a neighbouring antistar.

A number of indirect arguments have been presented in order to prove that there cannot be antimatter in our galaxy. We shall discuss them in VI.2.5.

VI.2.4.2. *Cometary Reservoir*

If half the stars in our galaxy consist of antimatter, we have to investigate what happens if a koinostar and an antistar pass close to each other. The probability of a collision between the two stars is very small but finite. If both stars are surrounded by planetary systems (including cometary reservoirs) like our Sun, there is a somewhat larger probability for a collision between a star and a planet of the opposite kind, but a much more likely collision would involve the comets in the reservoir. As these are believed to contain 10^{10}–10^{11} comets, there is a considerable chance that a comet will collide with a star of opposite kind.

During the close approach of two solar systems, the Leidenfrost layer separating the two different kinds of plasma regions may move in such a way as to allow a large number of comets to be situated in a thin plasma of the opposite kind. This plasma must hit the comets, but as an elementary calculation shows, no very conspicuous phenomena can be expected from this effect, unless the plasma has a density which is many orders of magnitude larger than normal interstellar plasmas.

If two stars of the same or opposite kind of matter pass each other at such a distance that neither of them penetrates the cometary reservoir of the other, the gravitational perturbation alone may still be large enough to eject some comets from the reservoir into interstellar space. One estimate of the rate of destruction of the Oort cometary reservoir by close stellar interactions has been made by Nezhinskij (1972) in which cumulative dispersion places a lower limit of the order of 10^9 years for the half-life of the cometary system, or the same order as the age of the solar system. Hence we should expect that there are a certain number of errant comets and that a star can be hit by a comet (perhaps of opposite kind of matter) even when it does not penetrate the cometary reservoir of another star.

VI.2.5. OBJECTIONS TO THE EXISTENCE OF ANTIMATTER

Any serious discussion of such a drastic revision of cosmic physics which the acceptance of the antimatter concept would naturally necessitate, is bound to meet with strong resistance from advocates of ingrained old theories. A number of more specific objections have been made, most recently by Steigman (1976a, b).

When Klein first discussed the existence of antimatter, it was claimed that from the measured upper limit of cosmic γ-radiation only an extremely small fraction of the matter in our galaxy could consist of antimatter. This conclusion was model dependent: the authors assumed that koino-matter and antimatter must necessarily form a homogeneous mixture. What has been said about homogeneous models in II.9, II.10 and IV.9 shows that this objection is not valid.

Another objection is that the energy source of an abundantly emitting object cannot be annihilation because very little or none of the hard annihilation γ-rays are observed. This conclusion is again model dependent. If the annihilation is produced by a solid (or gaseous) object falling on a star of opposite kind of matter, the annihilation takes place at the interphase, and there may be enough matter to absorb the γ-rays before they have escaped. See VI.3.3 and Figure VI.8.

Furthermore, the absence of a 0.5 MeV γ-line is cited as an objection to the existence of antimatter. This again is model dependent. The 0.5 MeV line should be emitted only when the electron–positron gas has cooled down to non-relativistic energies, which may be a rare case except in stellar atmospheres (IV.9.3). Annihilation of electrons–positrons with large kinetic energy will give a continuous γ-spectrum.

Another model dependent objection concerns the absence of anti-particles in cosmic rays of low and medium energies. This objection is based on the *assumption* that the magnetic field in the transplanetary region allows a free transport of cosmic rays in and out of the heliosphere. The development of the heliospheric model of III.4 indicates that this assumption is not valid.

Rogers and Thompson (1980) have recently analyzed the objections against antimatter – including those raised by Steigman – and have shown that none of them is crucial. They give a number of reasons why the antimatter is 'coy' and does not give an easily recognizable signature, but this is no reason why the universe could not be symmetric. In this chapter we will use several of their results.

VI.3. Annihilation as a Source of Energy

VI.3.1. ARGUMENT FOR THE EXISTENCE OF ANTIMATTER

Next to the rest mass energy the largest quantity of energy in the metagalaxy is the kinetic energy of the Hubble expansion. Depending on what model we choose, this amounts to about 5–20% of the rest mass energy (see Table VI.1).

This is obviously too large to be accounted for by ordinary nuclear reactions. This leaves us with *annihilation as the only possible source of energy* if we want to account for it by ordinary laws of physics. It should be observed that the Hubble expansion *cannot* be produced by an *explosion* in the singular point of the big bang cosmology (as is often erroneously stated in attempts to make the big bang cosmology popularly

Table VI:1

Hubble radius $R_H = 10^{26}$ m

Hubble time $t_H = R_H/c = 10^{10}$ y

Hubble volume $V_H = \frac{4\pi}{3} R_H^3 = 4.2 \times 10^{78}$ m^3

Density $\rho = 2 \times 10^{-29}$ kg m^{-3}

Galaxies are supposed to have:

$R_G = 10^{20}$ m, $M_G = 2 \times 10^{40}$ kg, $\rho = 5 \times 10^{-21}$ kg m^{-3}

β_{eff}	0.4	0.5	0.6	
z_{eff}	053	073	1	
Effective radius of metagalaxy $R_{eff} = R_H \beta_{eff}$	4	5	6	10^{25} m
Rest mass $M_M = 8.4 \times 10^{49} \beta_{eff}^3$	5.38	10.5	16.5	10^{48} kg
$N_\odot^M = \frac{M_M}{M_\odot} = 4.2 \times 10^{16} \beta_{eff}^3$	2.70	5.28	8.24	10^{15}
$N_G^M = \frac{M_M}{M_G} = 4.2 \times 10^{6} \beta_{eff}^3$	2.70	5.28	8.24	10^{5}
Rest mass energy $W_M = 7.35 \times 10^{66} \beta_{eff}^3$	483	944	1489	10^{63} J
Kinetic energy $W_k = W_M\, 0.25$ $B^2 = 2 \times 10^{66} \beta_{eff}^5$	19.3	59.0	141	10^{63} J
Microwave energy $W_\mu = 6 \times 10^{64} \beta_{eff}^3$	3.84	7.50	11.8	10^{63} J
Compare: Gravitational energy of metagalaxy $\sim 10^{62}$				
Density from de Vaucouleurs (2.2) $\rho =$	1.79	1.22	–	10^{-29} kg m^{-3}

accepted) because the Friedman model is homogeneous and there is no pressure gradient which can throw out matter. Instead the Hubble velocities are *postulated*.

Even aside from the Hubble expansion, there are several arguments for annihilation. During the last decade a number of celestial objects have been discovered in which the release of energy is too large to be accounted for by nuclear reactions. Without introducing new laws of physics, there are only two possible sources of energy; viz., annihilation and gravitational energy (IV.10.1). Quite a few theories have been proposed in which annihilation is supposed to deliver the energy (see for example, Alfvén, 1965, 1971, 1977; Elvius, 1970; Omnès, 1971) but in the recent discussion interest is so completely focused on gravitation as a source of energy that the annihilation theories are usually not even mentioned. In the preceding sections, it has been shown that the annihilation taboo is not rationally motivated.

We shall first give a brief survey of the situation and later (in VI.3.6 and VI.3.7) discuss some qualitative models for release of annihilation energy.

VI.3.2. SIMILARITY OF ELECTROMAGNETIC RADIATION FROM THE TWO KINDS OF MATTER

The kind of matter which a celestial body consists of cannot be determined by a study of the electromagnetic radiation it emits. The sign of some magnetic effects (circular Zeeman effect and Faraday rotation) depend not only the kind of matter but also on the sign of the magnetic field. However, as there is no independent way of measuring the sign of the magnetic field no conclusions can be drawn from the radiation about the kind of matter.

VI.3.3. RADIATION FROM ANNIHILATION PROCESSES

Mixing of the two kinds of matter produces *neutrinos, γ-rays, and relativistic electrons–positrons* (IV.9). It has been claimed that if large quantities of antimatter exist in the universe such radiation must be produced in quantities above the measured levels (Steigman, 1976a, b). This conclusion is not correct (Rogers and Thompson, 1980). Certainly, if we accept the usual picture of interstellar space as filled with a rather homogeneous plasma in turbulent motion we can exclude the existence of appreciable quantities of antimatter in our galaxy. However as we have seen in the first four chapters, recent magnetospheric–heliospheric observations compel us to change this view in a drastic way.

VI.3.4. LEIDENFROST LAYERS AND ANNIHILATION AT THE CELL WALLS

As we have seen in II.10 a number of interfaces have been discovered which separate regions of different magnetization, density, temperature, electron velocity distribution and even chemical composition. Examples are: the magnetopause and magnetotail sheets, the heliospheric equatorial sheet (earlier erroneously referred to as 'sector structure') and similar sheets in the magnetospheres of Jupiter, Saturn and Venus and probably sheets in the cometary tails also.

These sheets are caused by electric surface currents. They are sometimes very thin (down to a few times the ion Larmor radius). It is almost impossible to detect them from a distance. A spacecraft usually sees no indications of such a sheet until it actually passes through it.

As already mentioned, it is unlikely that cosmic plasmas have such properties only in those regions which are accessible to spacecraft diagnostics. Hence, it seems legitimate to assume that in general, space has a '*cellular structure*'. A revision of our concept of the properties of interstellar (and intergalactic) space is an inevitable constituent in the thorough revision of the theory of astrophysical plasmas which necessarily will follow from recent magnetospheric discoveries (Fälthammar *et al.*, 1978).

The new picture of the properties of space voids the objections against the evidence of antimatter within our galaxy. We are not in conflict with any observational facts if we assume that there are cells containing antimatter adjacent to cells with koino-matter and separated from these by Leidenfrost layers (VI.2.2). In two papers, Lehnert (1977, 1978) (see Figure VI.4) has studied such layers theoretically. In his first paper he assumes a homogeneous magnetic field parallel to the interface, in the second a combination between a Leidenfrost layer and a current sheet of the type observed in the magnetopause. As we know from observations that the current sheets in the magnetosphere–heliosphere have a high degree of stability, we can expect that even when combined with a Leidenfrost layer the sheet should be stable. As an example, suppose that the solar wind consisted of antimatter. When reaching the magnetopause, it would be deflected in the same way as the present solar wind (see III.6.7 and Figure III.1(c)) and very little antimatter would penetrate through the magnetopause.

VI.3.5. ANNIHILATION IN COSMIC CLOUDS

If the universe is symmetric with regard to koino-matter–antimatter, or if there at least exists a considerable quantity of antimatter, there are a number of situations in which annihilation may be important.

The first case we should consider is an encounter between cosmic clouds of opposite matter.

Annihilation processes produce a repulsion between the clouds and, according to recent investigations by Rogers and Thompson (1980), lead to an establishment of more or less stationary Leidenfrost layers of essentially the types studied by Lehnert. The total radiation from such a layer is calculated and only under extraordinary circumstances does it lead to the emission of a measurable quantity of γ-rays.

VI.3.6. BODIES FALLING INTO A STAR OF OPPOSITE KIND OF MATTER

Our second example refers to the fall of a small body (e.g., of the size of an asteroid or a cometary nucleus, typically 10^{14}–10^{16} kg) into a star. When the solid body hits the photosphere of the star, a burnout takes place in a few minutes. This seems to account for the γ-ray and X-ray bursts (Sofia and Van Horn, 1974, 1975; Vincent, 1976). The time constants of the rapid variations in γ-ray luminosity (see Figure VI.6) as well as the total emitted energy are of the expected order of magnitude, and the frequency of the events is reconcilable with the expected collision frequency.

The size spectrum of most groups of celestial bodies shows a rapid increase in numbers with decreasing size (e.g., Dohnanyi, 1976). Hence we should expect a large number of very small γ-ray bursts. Such bursts have actually been observed, mainly from balloons (see, for example, White *et al.*, 1978).

Going to larger bodies we should remember that in the solar system there are supposed to be 10^{11} comets but only 10^4 observable asteroids. The number of bodies of the size of satellites and planets is much smaller. If we take this distribution as representative of

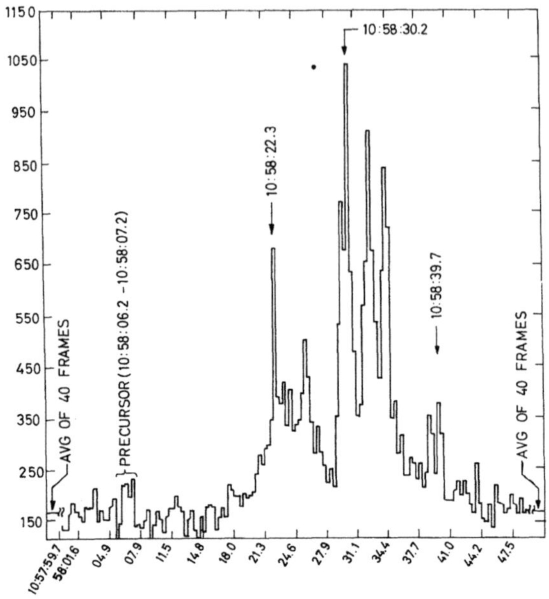

Fig. VI.6. Typical emission in a γ-ray burst. Total duration 50 s.

the size of distribution of bodies in our galaxy with which a randomly moving star is likely to collide, we should expect that with increasing mass of the impacting body the number of events should decrease rapidly. At the same time the quantity of emitted γ-rays should increase so that the γ-ray bursts become more conspicuous. However, in the ordinary γ-ray burst the γ-ray emission is already limited by absorption in the gas cloud produced by the evaporation of the small body (Thompson, 1978a) and most of the observed γ-rays are due to secondary processes in this cloud. The size of the evaporated cloud will increase with the size of the impacting body and, moreover, the burn-out will be delayed. Hence, the burn-out ultimately takes place under a cover of massive layers of the upper photosphere, so that there will be a saturation in the size of the γ-burst (Figure VI.7).

The relativistic e^+e^- gas which also is produced by annihilation is not subject to the same saturation. In fact, as will be discussed in detail later, it is likely to be emitted, even from large depths, in the form of jets or bubbles, which expand in the surroundings of the celestial body. In the presence of magnetic fields this component will emit synchrotron radiation, detectable at radio wavelengths and perhaps also at much shorter wavelengths. Hence, we should expect that the γ-ray burst should be accompanied by radio bursts, and also that there should be basically similar much more luminous radio bursts occurring less frequently.

The neutrinos cannot, of course, be appreciably absorbed but as our recorders are not yet sensitive enough we cannot obtain any information from them (Rogers and Thompson, 1980).

In the next section, we shall treat two cases of collisions between condensed bodies,

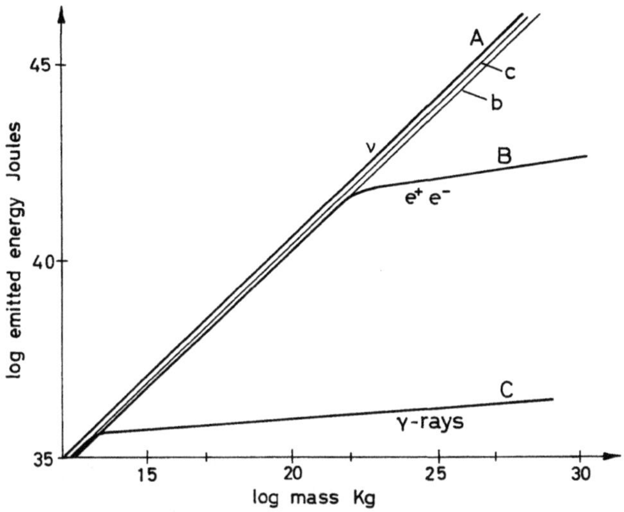

Fig. VI.7. Energy emission W when a condensed body with mass m annihilates in a star. Curve (A): $W = mc^2$. Neutrino energy and also total non-neutrino energy. Represents upper limit to total electromagnetic radiation (including light) from ambistar. (b) $W = \frac{1}{3}mc^2$. Energy released as e^+e^- relativistic gas. (B) Most of the e^+e^- is emitted but self-annihilation gives saturation at $M \sim 10^{22}$ kg. The curve gives the upper limit to emitted synchrotron radiation and also to the contribution to the continuous X-ray background radiation. (c) $W = \frac{2}{3}mv^2$. Energy released as γ-rays. (C) Because of absorption in the condensed body and the cloud produced when it evaporates, the emitted γ-radiation saturates, in part already when $m \sim 10^{13}$ kg.

viz., when a star of solar dimensions is hit by a 'medium size' body, say of terrestrial mass, and when the impacting body is of stellar mass.

VI.3.7. MODEL OF AN AMBISTAR

Because our basic knowledge of how koino-matter and antimatter interact is still rudimentary, it seems impossible to make a realistic model of an ambistar. What we may be able to do is to discuss a qualitative highly simplified model with the purpose of clarifying what major processes should be taken into consideration. This was done in a recent paper (Alfvén, 1979a). It was supposed that a star of solar dimensions is hit by a much smaller (Model I) or comparably sized (Model II) body of the opposite kind of matter. The observational results which were used for comparison were mostly taken from Burbidge *et al.* (1974) and a number of papers in *Physica Scripta* (1978).

VI.3.7.1. *Ambistar Model I*

In this model it is assumed that a star is impacted by a body of the opposite kind of matter which is much smaller but still large enough not to burn up immediately (which probably means $\gg 10^{20}$ kg). After a very violent but brief initial phase (order of minutes) the matter of the impacting body will be a part of a composite star ('ambistar') containing both kinds of matter separated by a Leidenfrost layer. After a transient period (probably less than 100 years) the flow of energy inside the ambistar may reach a quasi-stationary state. In the separating layer annihilation will produce neutrinos, γ-rays, and

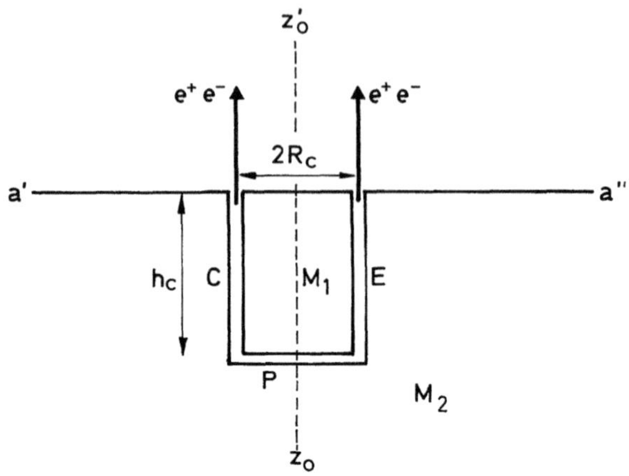

Fig. VI.8. Ambistar Model I. Near the surface $a'a''$ of a star of one kind of matter there is a volume of matter of the opposite kind, limited by a circular surface P and a cylindrical surface C. The axis of symmetry is z_0z_0'. The two kinds of matter are separated by Leidenfrost layers. A relativistic gas of e^+e^- (10^8 eV) is ejected, essentially perpendicular to the stellar surface. In case the height of the cylinder is so large that the mass density exceeds $\sim 10^3$ kg m^{-2}, most of the γ-rays are absorbed. The intense energy release makes the surface region near the cylinder very hot, causing a very strong emission of light.

a relativistic e^+e^- gas. Because of the high temperature this region is likely to stay at the surface of the ambistar. Only a highly simplified model is discussed.

The photosphere of the ambistar is represented by a plane $a'a''$ (see Figure VI.8). The matter M_1 originating from the infalling body is supposed to be confined inside a circular cylinder with the radius R_c and the height h_c. It is separated from the rest of the star M_2 by a circular Leidenfrost layer P called the *production region* where all the annihilation is supposed to take place, and the cylindrical shell with thickness Δ is called the *exhaust channel E*. The annihilation which takes place in this is neglected.

The star is assumed to rotate and to be magnetized and it is argued that the cylinder of opposite matter will find an equilibrium position near the axis of rotation.

The model is worked out in some detail and is found to have the following properties. The annihilation produces neutrinos, which will leave the ambistar without causing any important phenomena. Further, it produces γ-radiation which is mainly absorbed in the close neighborhood of the production region, where it will cause an intense heating and also produce secondary e^+ and e^-. As the annihilation takes place mainly in the production region which is screened from space by a thick layer of matter, a *rather small emission of* γ-rays can be expected. The e^+e^- produced in this way, and especially those directly produced at the annihilation will also cause intense heating but due to the magnetic field they will form gas clouds which will escape into space through the exhaust channels. A rather large fraction of them will be emitted as a more or less well-defined beam in the vicinity of the axis. In the stellar magnetic field they will produce an *intense synchrotron radiation*. The strong heating of the plasma on both sides of the exhaust channel will also cause an *emission of both koino-matter and antimatter* in a cone around the axis. Calculations show that outputs of, for example, 10^{36} W are reasonable.

VI.3.7.2. *Stellar Collisions*

Next we consider the case when the two colliding bodies of opposite matter have comparable masses.

We shall first discuss the moment of collision. In some respects the collision between two stars of opposite kinds of matter is similar to the collision between two stars of the same kind of matter. Because of its importance for the creation of very massive stars, especially in galactic cores, this problem has attracted considerable interest and has been treated by Spitzer and Saslaw (1966), Mathis (1967), Colgate (1967), Sanders (1970) and Seidl and Cameron (1972) (For a brief summary, see Saslaw (1975)).

Depending on their models the different authors reach somewhat conflicting results. It is obvious that the collision is highly inelastic and that there is an appreciable probability that two stars can collide in such a way that they form a combined star with a mass which is not very much smaller than the sum of their masses. The combined star is strongly heated resulting in a great enhancement of nuclear energy release and the generation of violent oscillations. However, as the time of collision is very short ($\sim 10^3$ s) most of the oscillations will decay rather rapidly, compared to the time scale which is relevant for our general considerations.

If two stars of opposite matter collide it seems reasonable that the very brief collision phase is similar. As annihilation is a surface phenomenon, primarily confined to a layer only a few mean free paths thick, the release of annihilation energy will not necessarily be large enough to affect the kinetics of the collision (in an analogy to the release of nuclear energy).

VI.3.7.3. *Ambistar Model II*

What situation will be established after the transient effects have decayed, is not very easy to calculate with any high degree of certainty. For reasons discussed in Alfvén (1979a), a reasonable model (Figure VI.9) should be based on the assumption that the

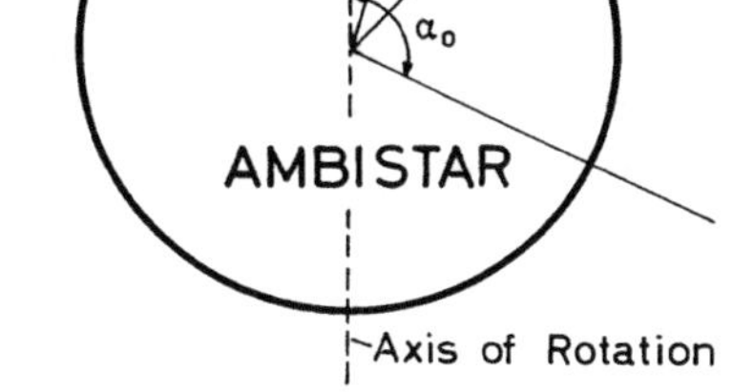

Fig. VI.9. Highly simplified model of an ambistar. Plasma, e^+e^- gas and light is emitted from a very hot polar cap ($\alpha < \alpha_1$). The light is redshifted when seen by an observer at $\alpha < \alpha_0$, and blueshifted for $\alpha > \alpha_0$. The ambistar is identified as a QSO only if its highly luminous polar cap is visible, with the consequence that QSOs always are redshifted. If blueshifted, the ambistar appears as a rather normal star.

properties we have derived from Model I are essentially also valid for an ambistar in which the two components are of comparable mass. Hence, the emission is spin stabilized and consists of e^+e^- plasma, synchrotron radiation, and light. The recoil of these particles produces a spin stabilized rocket acceleration of the ambistar. If the initial masses of the two kinds of matter are almost equal and the same amount of plasma of both components is emitted an almost complete burnout can be achieved. The rocket acceleration will then push the increasingly small mass of the ambistar to very large velocities, and *in principle any value of the redshift z is attainable*.

It should be stressed that such extreme accelerations should be rather rare. When two stars of opposite matter collide and form an ambistar, it is only when the masses are almost equal that a very large acceleration is likely to occur. Hence, *all new-born QSO's and many of the old QSO's should have very small velocities in relation to their mother galaxies*.

VI.3.7.4. *Observable Properties of an Ambistar*

The ambistar model which we summarize here is highly asymmetric. This means that standard stellar models are not applicable. Its observable properties depend on the angle between the observer and the axis (see Figure VI.9). Around the axis there is a cone with opening angle α_0 in which there is an intense light emission so that if an observer is located at an angle $\alpha < \alpha_0$ he will observe a luminosity orders of magnitude larger than if he is at a larger angle ($\alpha > \alpha_0$). Consequently, to observe such an enormous luminosity and to identify it with a QSO, it has to be seen at an angle $\alpha < \alpha_0$. Should $\alpha > \alpha_0$, the ambistar will be seen as an object with a more normal luminosity – brighter than a normal star of similar size but not at all comparable to a QSO.

VI.3.7.5. *Blueshifts*

One of the important arguments against an *intrinsic acceleration* of QSO's to large velocities – as an alternative to the conventional 'cosmological velocity' doctrine – is that very few if any QSO's with blueshifts are observed. The ambistar Model II explains why. If we observe an ambistar, it may just as well be blueshifted as redshifted, but we will not identify it as a QSO unless we observe it from the backside where its rocket is burning. If the acceleration is spin stabilized we will observe it preferentially as redshifted.

VI.3.7.6. *QSO Scenario*

Assuming annihilation to be the energy source of the QSO's and adopting Ambistar Model II we arrive at the following scenario for these objects (Alfvén, 1979a).

(1) *The origin of a QSO* is a collision between a koino-star and an antistar. The place of birth is likely to be in the dense core of a galaxy because stellar collisions are most frequent there.

(2) The *collision* is not very different from the collision between two stars of the same kind of matter. Unfortunately, this process is not very well understood, but existing models seem to indicate that if the collisional velocity is not excessive, a composite star will be formed after a very violent period of minutes or hours. Its mass is probably not less than 90% of the sum of the masses. As the kinetic energy of the impact is added to the sum of the internal energies of the colliding stars, the composite star will initially be very hot and in a state of violent oscillations. The high temperature will speed up the

nuclear reactions in the core, but the nuclear energy release during the short collision period will be small compared to the kinetic energy and will not affect the collisional process appreciably. After a time of the order of years or more the composite star will settle down to the properties of a normal star of the same mass.

(3) These results may be applicable to the brief collisional period during which an ambistar is formed. Annihilation, confined to a very thin boundary layer between the two kinds of matter, is not likely to be large enough to change the collisional process. For reasons given in VI.3.6, we give a tentative description of an ambistar by two semi-empirical *ambistar models*, the first one referring to impacting bodies with very large difference in mass, the second one to bodies with comparable masses. The latter model should describe a typical QSO.

(4) The two kinds of matter will be separated by a Leidenfrost layer inside the ambistar in which an intense annihilation takes place. The annihilation gives rise to neutrinos which are escaping rapidly, to γ-rays, which are largely absorbed inside the ambistar, and to a relativistic e^+e^- gas which will be largely emitted, in a direction parallel (or anti-parallel) to the rotational axis. Intense heating, mostly from the absorbed γ-rays, will occur in regions around the axis. The hottest region may be a circle around the axis (as in Model I) or the whole *polar cap*. There will be a plasma emission, most intense above the hottest region which we assume extends to an angle α_1 from the pole. As in the case of the solar wind the ejection velocity will vary, with the result that density variations build up, eventually resulting in the formation of a number of discrete clouds.

(5) Under certain circumstances, the recoil force of the emitted e^+e^- gas, light emission and plasma emission may give an efficient *rocket acceleration* of the ambistar. The 'exhaust gases' will be confined to the angle α_1 from the axis (Figure VI.9).

(6) Many QSO's will not be accelerated very much. Hence they will have the same redshift as their mother galaxy.

(7) Under special conditions, including that the two colliding stars have almost the same mass, the rocket effect may accelerate the ambistar, possibly up to close to the velocity of light. As a galactic core usually has the dimension of 10^{20} m, an ambistar travelling with a velocity $\sim 3 \times 10^8$ m s^{-1} is likely to remain inside it for a time of the order of $10^{20} (3 \times 10^8 \times 3 \times 10^7)^{-1} = 10^4$ years. An escape from the galactic core is possible if the birthplace is in the outer region of it and its lifetime is long enough.

(8) If the light emission takes place in a beam, the observed properties of an ambistar depend on the location of the observer in relation to the axis. Hence *there should exist blueshifted ambistars*, but they will not look very much different than ordinary stars. Due to the annihilation heating their luminosity may be, perhaps, 10 or 100 times larger than that of an ordinary star but *they will not be identified with a QSO* which normally is $\gg 10^6$ times more luminous. (Perhaps 10^{35} W compared to 4×10^{26} W for the Sun.)

Hence, the *absence of observed blueshifts is not an argument against our local acceleration model*.

(9) According to Figure VI.9 a QSO should appear as a strongly luminous core associated with a jet (although we cannot be sure how easily the jet can be observed). There are quite a few QSO's observed to be associated with jets (see, for example, Burbidge *et al.*, 1978; Readhead *et al.*, 1978; Pearson *et al.*, 1980).

(10) The energy release of an observed QSO has so far always been calculated under

the assumption of an isotropic emission of radiation. According to our model the emission is highly anisotropic and the usual values of energy release should be reduced by perhaps one order of magnitude. Further, because QSO's may be located at distances much smaller than the 'cosmological' distances, the energy release figures may be revised downwards by still one or in some cases several orders of magnitude.

(11) A QSO will become an ordinary star when the mass of the smaller component is completely annihilated. The lifetime of a QSO cannot exceed $t = 2M_s c^2/P_{QSO}$, where M_s is the mass of the small star and P_{QSO} is the power generated from annihilation. As a rather arbitrary numerical example with $M = 10^{30}$ kg and $P_{QSO} = 10^{35}$ W we find $t = 2 \times 10^{12}$ s $= 10^5$ years.

As the total number of known QSO's is of the order 10^3, there must be more than 10^{-2} stellar collisions per year leading to the formation of observable QSO's.

VI.3.7.7. *Continuous X-ray Background Radiation*

The relativistic e^+e^- gas which is emitted in these events will expand out in space, and will eventually fill intergalactic space. As shown by Carlqvist and Laurent (1976a, b) it should be possible to observe it there because starlight shining on it will produce X-rays via the inverse Compton effect. In this way it explains the observed continuous X-ray background radiation without any *ad hoc* assumptions. The shape of the theoretical spectrum agrees well with the observed spectrum in the energy range ≈ 1–100 keV, and the intensity is of an acceptable order of magnitude.

Objections to this interpretation have been raised by Steigman (1976a) but Carlqvist and Laurent (1976b) have demonstrated that they are not valid (see also Rogers and Thompson, 1980).

VI.4. Hubble Expansion in a Euclidean Space

VI.4.1. NON-COSMOLOGICAL REDSHIFTS OF SOME QSO'S

The conclusion that intrinsic effects like a rocket mechanism in some special cases can accelerate ambistars to velocities approaching c is supported by a number of observational data. For a long time Arp (see for instance Arp, 1966, 1977, 1978) has claimed that QSO's with very high redshifts are so often found close to galaxies with much smaller redshifts, that they must be genetically connected, in spite of the large difference in redshifts. Burbidge (1973, 1979) has summarized the results of Arp and others. He concludes that in many cases a QSO close to a galaxy has the same redshift as the galaxy, but that there also are many cases when a QSO has a much larger redshift than the adjacent galaxy. In some cases observed by Arp the velocity of the QSO's *with reference to the galaxy* in which it was born must exceed $\beta = v/c = 0.8$. A body with this velocity has a kinetic energy which is a fraction $(1 - \beta^2)^{-1/2} - 1 = 0.67$ of its rest mass energy. The only reasonable explanation for such high kinetic energies seems to be that the body has emitted radiation or/and plasma unidirectionally and has been accelerated by the resulting rocket recoil of a type similar to that which we have discussed in VI.3.7. The energy needed for such an acceleration should be compared with the energy release which in any case must be assumed in order to account for the QSO's very large luminosity. If the two components of the ambistar have almost equal mass a complete burnout will result in a kinetic energy which is a large fraction of the rest mass energy.

We conclude that – at least in some cases – the radiated energy may be of the same order of magnitude as is needed to accelerate a QSO of for example one solar mass up to very high z-values.

Hence *without any assumptions other than that the energy source of a QSO is intrinsic, we find that the observed radiated energy and the energy required for rocket acceleration to relativistic velocities are of comparable order of magnitude.* This conclusion could be checked observationally by comparing the energy dissipation of a QSO with its velocity (with reasonable assumptions about its lifetime and mass).

The essentially semi-empirical model which we have discussed above seems to give a reasonable picture of several of the observed properties. It is concluded that only annihilation can provide the energy which is required.

Again it should be stressed that very large accelerations of a QSO take place only under certain conditions. *Many QSOs*, for example those who are just born and those where the mass ratio of the two colliding stars is far from unity, will *have a velocity which is almost the same as that of their mother galaxy*.

Hence, in summary:

(1) accepting that many QSOs, *but not all of them*, are at 'non-cosmological' distances,

(2) accepting that the energy source must be intrinsic,

(3) assuming that the energy is emitted unidirectionally (which is necessary in order to explain the absence of blueshift),

(4) assuming that the mass of the QSOs is comparable to the solar mass,

we conclude that some QSOs can be accelerated up to relativistic velocities.

It is important to stress that the cosmological conclusions in this chapter are *not based* on the existence of antimatter. Any other intrinsic energy source is acceptable if it satisfies the mentioned conditions.

VI.4.2. CONSEQUENCES OF NON-COSMOLOGICAL INTERPRETATION OF QSO REDSHIFTS

If we accept that some QSOs have 'non-cosmological' redshifts, i.e., that they are accelerated by an intrinsic release of energy, this has far-reaching consequences for cosmology. Not only the QSOs themselves but also the clouds emitted from them during their acceleration (and sometimes observed as absorbers of their emission sometimes as jets), should be explained as due to an intrinsic release of energy in the QSO. Hence those QSOs which have very large z-value should not be included in the analysis of the general Hubble expansion of galaxies. This means that the large scale structure of the metagalaxy must be derived from the measurements of objects which can be confidently identified as structures unrelated to QSOs.

The largest redshifts which with certainty have been identified with galaxies are $z = 0.806$, 0.811, and 0.890 (Smith *et al.*, 1979), which means $\beta = 0.53$, 0.53, and 0.57.

de Vaucouleurs defines the effective radius R_{eff} of an assembly like a cluster or supercluster by the condition that for half of its members the projected radius should be less than R_{eff}. Hence even if we accept the assumption that the Hubble parameter is constant out to the Hubble limit R_H, we cannot put R_{eff} of the metagalaxy equal to the Hubble radius, because this would mean that a large number of galaxies should receed with $v > c$. If instead we put $\beta_{\text{eff}} = R_{\text{eff}}/R_H = 0.5$ (which means $z_{\text{eff}} = 0.73$) we would still have more matter outside than inside R_{eff}. Hence the observed β-values are well reconcilable with $R_{eff} = 0.5 R_H$ (and probably also with $\beta_{eff} = 0.4 R_H$ $(z = 0.52)$ or perhaps even less.

We shall use $\beta_{eff} = 0.5$ $(z = 0.73)$ as a standard model. This allows quite a few galaxies to have z-values as high as $z = 1.0$, and some even higher. Not until we can be sure that there is more matter outside $z = 0.73$ than inside do we have any reason to increase our value of R_{eff}. (It should be observed that a higher value of R_{eff}, for example $0.6 R_H$ $(z = 1.0)$ does not imply any major modifications of our conclusions.)

VI.5. A Model for the Evolution of the Metagalaxy

The implications of these assumptions are shown in Table VI.1 (p. 137). The values of R_H and t_H (Hubble time) derive from de Vaucouleurs. The density ρ is an upper limit to the values for $\beta = 0.4$, 0.5 and 0.6. The kinetic energy W_k depends on whether we assume a constant density or a density distribution inside the metagalaxy. The value of 0.25 of the coefficient in the kinetic energy formula is a compromise. The microwave energy is based on a blackbody radiation of 3 K, giving an energy density of $w = 1.53 \times 10^{-16}$ J m^{-3}.

VI.5.1. THE PROTO-METAGALAXY

It is of interest to investigate how far backwards in time we can reconstruct the evolution of the metagalaxy from the present observations, *independent of all cosmological theories*.

From the Hubble expansion we can conclude that earlier the metagalaxy was smaller than now. However, because observers disagree about the value and the constancy of H_0 we cannot conclude with certainty how large its minimum size was. Figure VI.2 shows that even observations which claim to prove its constancy are reconcilable with a minimum size of 0.1 of its present size. The general disagreement about its absolute value and its anisotropy indicate that even a larger minimum size can not be excluded. In the following we will assume that in the past the metagalaxy was in a state called the *proto-metagalaxy* when its size was about 0.1 of its present size.

Before this time the metagalaxy may have been (a) much smaller (as the big bang hypothesis claims), (b) stationary in size, at least for some time or (c) contracting from a larger size to the size $0.1 R_H$ (as Klein postulated).

In order to give a better basis for discussing these three options we should first reconstruct the state of the proto-metagalaxy, a state of the metagalaxy about $0.9 t_H$ ago when its size was about

$$R_{PM} = a \cdot R_H \tag{9}$$

with $a = 0.1$.

As we have found in VI.1.4.2 we can be pretty sure that for this reconstruction we can use Euclidean geometry. With $\Omega < 10^{-3}$ as we have found in VI.1.3.3 the general relativity correction would never exceed about 1%. (Even with $\Omega < 0.06$ a Euclidean description would be legitimate but with a rather large relativistic correction.)

It is premature to try to give a detailed picture of the proto-metagalaxy. We shall not attempt to guess whether it consisted of galaxies or of a more even distribution of stars or clouds. A brief discussion of how it might have been formed is given in VI.5.4.

The only property which is so important that we must discuss it is whether its state was such as to generate the Hubble expansion.

VI.5.2. THE KINETIC ENERGY OF THE HUBBLE EXPANSION

The alternatives (b) and (c) imply that the proto-metagalaxy must be able to produce the kinetic energy of the Hubble expansion, which is the largest energy after the rest mass.

In the more popular presentations of the big bang model it is often claimed that the Hubble expansion is due to an 'explosion' but as in the Friedman model there is no pressure gradient, this is not correct. In reality the high velocities are postulated and to the extent they are 'explained' they are attributed to supernatural effects associated with an *ex nihilo* creation of the universe. As our approach is based on the assumption that no new laws of nature are needed, we cannot accept such an explanation. Instead we must determine in the usual scientific way from what phenomena the Hubble expansion obtains its energy.

As was pointed out in IV.10.1, the only sources of energy with a capacity large enough to explain this amount of kinetic energy, is gravitation and annihilation. Gravitation is, of course, responsible for a great many processes in astrophysics. None of these, however, can convert energy in the scale needed to accelerate the metagalactic masses to Hubble velocities. Attempts at explanations, capable of large scale conversion, often concern black holes.

However, the most intense release of energy we know is found in the QSOs. Hence without reference to any specific model we would suspect that the QSO type of energy release is the best candidate for an energy generation in the proto-galaxy.

Accepting the non-cosmological interpretation of the QSO velocities (VI.4) we have found that in these objects there must also be a mechanism which can accelerate large masses to the relativistic velocities of the QSOs and also of the ejected gas clouds. This may be the mechanism we need in order to generate the kinetic energy of the Hubble expansion.

A calculation shows that the required QSO activity is certainly much larger than the present one, but it does not seem to be unreasonably high. In fact, in a galaxy of, for example, 10^{10} solar masses, about one solar mass should be consumed per year (Alfvén, 1979b).

If the QSOs derive from stellar collisions, it is reasonable to assume them to be much more frequent in the proto-metagalaxy, the density of which is 10^3 larger than in the present metagalaxy.

VI.5.3. FORMATION OF THE PROTO-METAGALAXY

With the above discussion in mind, we can summarize some properties of the proto-metagalaxy. Possible numerical values are given in Table VI.2. Speculations about the other properties of the proto-metagalaxy are found in Alfvén (1979b). We come to the conclusion that the proto-metagalaxy has not necessarily originated from a big bang at a singular point. The alternatives (b) and (c) in VI.5.1 are also possible, because the kinetic energy of the Hubble expansion may have been supplied by an intense QSO activity (or QSO-like activity) when the metagalaxy had the size R_{PM}. The advantage of this assumption of the source of energy is that *it relies on a type of phenomenon which has been really observed.*

The problem of how the proto-metagalaxy formed, easily drags one into a jungle of cosmological speculation. To try to solve this problem is not the purpose of this

Table VI:2
Properties of proto-metagalaxy

Radius of protometagalaxy	$R_{PM} = 5 \times 10^{24}$ m
Mass (equals present mass increased by less than 50%)	$M_{PM} = 0.3\text{–}1.0 \times 10^{49}$ kg
Density	$\rho = 0.6\text{–}2 \times 10^{-26}$ kg m^{-3}
Escape velocity	$(2GM/R)^{1/2} = 1\text{–}2 \times 10^7$ m s^{-1} $\sim 0.05c$
Laplace–Schwarzschild limit	$R_{Sch} = 10^{22}$ m
Correction for general relativity	$\approx R_{Sch}/R_{PM} \approx 1\%$
Correction for special relativity (except for QSO related phenomena)	$\approx \beta^2 \approx 25\%$

monograph. In a Euclidean model time may be infinite and for every earlier period we explore, a new problem of still earlier states will appear. The further we go back in time the less we can say with any degree of certainty. Cosmology as a science should work towards clarifying the conditions in ever increasing regions of space and time but a discussion of a possible 'ultimate cause' belongs to philosophy or religion.

It should be remembered, however, that Klein assumed that the proto-metagalaxy was formed by infall of matter from an immensely large, but not infinite, sphere containing a homogeneous ambiplasma (mixture of koino-matter and antimatter). This view is reconcilable with our alternative (c) in VI.5.1, with the rather important reservation that Klein's model is homogeneous. We know now that most of the early homogeneous models in astrophysics are grossly misleading and must be replaced by inhomogeneous models. For our problem, this is important because an infall of matter–antimatter may take place in any period. Such a process can therefore be an integral part of the evolution of the metagalaxy during both very early and very late periods, including the present period when the dominating dynamic phenomenon is the Hubble expansion.

VI.5.4. OTHER COSMOLOGICAL PROBLEMS

There are a number of other important cosmological problems which a Euclidean approach to cosmology has to account for – sooner or later. Some of them may be solved by a straightforward application of the matter–antimatter symmetry. We have already discussed the *γ-ray bursts* (VI.3.6) and the *X-ray background radiation* (VI.3.7.7). These phenomena can only be explained in the big bang cosmology by a number of questionable *ad hoc* assumptions.

The *production of heavy elements* might take place in QSOs where the temperature varies from the normal temperatures of stellar interiors to extremely high temperatures in the regions close to the 50 MeV e^+e^- gas in the Leidenfrost layer. The present theories need both the moderately high temperatures in stellar interior and the extreme temperatures of the big bang and the supernova explosions. Both seem to be available in the QSOs.

An interesting problem is the generation of the *microwave radiation* in a hierarchical cosmology. As seen from Table VI.1 there is more than enough radiative energy to produce the microwave radiation, so the essential problem is to find a mechanism which can accomplish this conversion and give the rather high degree of isotropy. Of a number of conceivable mechanisms which have not yet been discussed seriously, we shall only mention one or two. The isotropy may be the result of the conversion taking place in some shell or cocoon-like structure. We know that nova and supernova explosions produce expanding shells around them. These consist of dust and gas, probably to a large extent in a molecular state. Similar shells surrounding galaxies have recently been detected by Malin and Carter (1980). It is as yet unknown whether they contain grains or molecules which may isotropize the microwave radiation. Whether there are still larger shells, e.g., surrounding galactic clusters or superclusters is not known.

The millimeter wavelength region is the least explored in the whole electromagnetic spectrum. Time is not ripe for a theory of the 3 K microwave background until the structure and chemical composition of such shells are better understood and until the properties of millimeter radiation have been thoroughly explored.

Another approach to the microwave background is tentatively suggested in the next section.

The advance of X-ray and γ-ray astronomy has demonstrated the existence of a large number of extremely energetic events. The energy source of these is often supposed to be 'black holes'. The mechanisms for energy emission from black holes are very complicated. It seems that, at least in many cases, annihilation is a more likely energy source.

VI.6. Other Metagalaxies

An interesting feature of a Euclidean cosmology is that there may be other metagalaxies in the vicinity of ours. A Euclidean cosmology does in principle accept an infinite Universe. (Certainly an open Friedman model is also infinite from a mathematical point of view, but as it does not allow anything to exist outside the present Hubble distance, it is finite from a physical point of view.) It is conceivable that a number of metagalaxies together may form a still higher order (which has been called a 'teragalaxy') in a hierarchical cosmology. If the microwave radiation is not enclosed in the metagalaxy but fills a large space with the same energy density 1.5×10^{-14} J m^{-3} $= 1.4 \times 10^{-31}$ kg m^{-3}, the Laplace–Schwarzschild limit is reached at $R = 3 \times 10^{28}$ m. Hence a teragalaxy of this dimension might be closed by the mass of the microwave radiation alone. In this highly hypothetical case, Euclidean geometry is not applicable to the whole universe but only to minor parts of it, like metagalaxies or possibly somewhat larger regions.

VI.7. Discussion

The big bang advocates often claim that a Euclidean cosmology must be necessarily 'homocentric' or 'pre-Copernican' and hence from a philosophical point of view 'distasteful'. This may be true if a homogeneous Euclidean model is accepted, but a hierarchical model gives a different picture – simply because it has no centre!

Further, the Copernican model left the Earth in a rather central position because our solar distance is only one tenth of the average distance of a planet. If we put the outer limit of the solar system at 40 AU, the earth is located in the innermost 10^{-5} part of it. If we include the cometary reservoir of 10^4–10^5 AU the fraction is reduced to 10^{-12}–10^{-15}. Hence in the language of the big bang advocates the Copernican system should be regarded 'distastefully homocentric'.

Moreover, our sun is about *ten times more distant from the galactic centre than an average star*. Our galaxy does not occupy a central position in our local group, which is peripheral in our cluster of galaxies, which again is far from the center of our supercluster. Hence even if it were true that a Euclidean model must give our very impressive supercluster a central position in our metagalaxy, this could not be brandished as a homocentric view. However, a Euclidean cosmology need not necessarily do even that, because *we know that our supercluster moves with a rather large velocity in relation to the microwave background*. Furthermore, the relative velocity might be zero if both the supercluster and the background radiation share the same Hubble expansion. It should be noted that the proto-metagalaxy has a column density which should give some coupling between matter and radiation.

Finally, the Euclidean cosmology questions whether our metagalaxy really is the whole universe – which the big bang advocates claim. This is quite the opposite to making the universe, the real infinite universe, more homocentric.

VI.8. Conclusions

In the first four chapters a survey has been given of the rather drastic revision of the views on cosmic plasmas, which has been caused by *in situ* measurements in space and by the translation of laboratory results to cosmic physics. Plasma phenomena should be described not only by magnetic field models but also by explicitly accounting for electric currents and the circuits in which these flow.

This revision of cosmic plasma physics cannot be limited to those regions which are accessible to *in situ* measurements but must – sooner or later – be extended to the astrophysics of more distant regions. Some examples of what this may imply have been given, including the problem of the formation of the solar system (in Chapter V).

In this chapter we have studied what consequences – if any – the new views may have for cosmology. Two important conclusions are drawn.

(1) The electric current formalism draws attention to the attraction between currents which has a tendency to produce inhomogeneities. If we add to this the fact that also gravitation has the same tendency, it is natural that *inhomogeneous models should dominate astrophysics*. Observations clearly confirm this. In fact homogeneous models are often misleading, even as first approximations.

The most important field of astrophysics where homogeneous models are generally accepted is cosmology; indeed the Friedman model of the Hubble expansion is *homogeneous*, and moreover *four-dimensional*. It is shown that there are no decisive observational facts making a model of this type necessary. The properties of our metagalaxy ('universe') are more adequately described by *inhomogeneous Euclidean* models of the same type as is generally used in other fields of astrophysics.

(2) Largely *independent* of the criticism of the big bang hypothesis, the possible matter–antimatter symmetry of the universe is discussed. The observational discoveries of the cellular structure of space and of the hierarchical density distribution in the metagalaxy seem to be favourable to this approach.

The analysis in this monograph can be considered as an attempt to see to what extent the observed cosmic phenomena can be understood without postulating any new basic laws of physics. It seems that so far no phenomena have been discovered which necessarily call for any new laws. The basic properties of a plasma seems to be the same everywhere, from the laboratory to the Hubble distance.

REFERENCES

Chapter I

CE refers to Alfvén and Fälthammar (1963).

Akasofu, S.-I. and Chapman, S.: 1972, *Solar-Terrestrial Physics,* Clarendon Press, Oxford.

Alfvén, H. and Fälthammar, C.-G.: 1963, *Cosmical Electrodynamics*, Int. Series of Monographs on Physics, Clarendon Press, Oxford.

Birkeland, K.: 1908, *The Norwegian Polaris Expedition 1902–1903*, Vol. 1, Sect. 1, Aschehoug, Oslo.

Block, L. P.: 1976, in D. J. Williams (ed.), *Physics of Solar Planetary Environments*, Proc. AGU Symp., Boulder, Col.

Dessler, A. J.: 1979, *Chapman Conference on Magnetospheric Boundary Layers*, Historical Review, Alpbach.

Fälthammar, C.-G.: 1974, *Space Sci. Rev.* **15**, 803.

Fälthammar, C.-G., Akasofu, S.-I., and Alfvén, H.: 1978, *Nature* **275**, 185.

Kadomtsev, B. B.: 1965, *Plasma Turbulence*, Academic Press, New York.

Krall, N. A. and Trivelpiece, A. W.: 1973, *Principles of Plasma Physics*, International Series in Pure and Applied Physics. McGraw-Hill, New York.

Mendis, D. A.: 1978, *The Moon and the Planets* **18**, 361.

Motley, R. W.: 1975, *Q Machines*, Academic Press, New York.

Nagoya University Institute of Plasma Phyics: 1977, Research Report, Workshop on Relation Between Laboratory and Space Plasmas, IPPJ-286, Nagoya, Japan.

Nagoya University Institute of Plasma Physics: 1979, Research Report, Second Workshop on Relation Between Laboratory and Space Plasmas with Particular Reference to the Heliosphere, IPPJ-365, Nagoya, Japan.

Potemra, T. A.: 1978, *Astrophys. Space Sci.* **58**, 207.

Siscoe, G. L.: 1979, in B. Battrick (ed.), *Magnetospheric Boundary Layers*, Proc. Chapman Conf., Alpbach, June 1979, European Space Agency SP-148, Noordwijk, Holland.

Wurm, K.: 1963, B. M. Middlehurst and G. P. Kuiper (eds.), *The Moon, Meteorites and Comets*, Univ. of Chicago Press, Chicago.

Chapter II

CE refers to Alfvén and Fälthammar (1963).

Akasofu, S. -I., Lee, L. C., and Gray, P. C.: 1979, 'A model of the interplanetary magnetic field configuration', (private communication).

Alfvén, H.: 1977, *Rev. Geophys. Space Phys.* **15**, 271.

Alfvén, H.: 1979, in B. Battrick (ed.), *Magnetospheric Boundary Layers*, Proc. Chapman Conf., Alpbach, June 1979, European Space Agency SP-148, Noordwijk, Holland.

Alfvén, H. and Fälthammar, C.-G.: 1963, *Cosmical Electrodynamics*, Int. Series of Monographs on Physics. Clarendon Press, Oxford.

Andersson, D., Babic, M., Sandahl, S., and Torvén, S.: 1969, *Proc. 9th Int. Conf. on Phen. in Ionized Gases*, Editura Academii Republicci Socialista Romania, Bucharest.

Baker, K. D., Block, L. P., Kist, R., Kampa, W., Singh, N., and Thiemann, H.: 1980, (submitted to *J. Plasma Phys.*).

Bennett, W. H.: 1934, *Phys. Rev.* **45**, 890.

Block, L. P.: 1972, *Cosmic Electrodyn.* **3**, 349.

Block, L. P.: 1978, *Astrophys. Space Sci.* **55**, 59.

Block, L. P. and Fälthammar, C.-G.: 1968, *J. Geophys. Res.* **73**, 4807.
Boström, R.: 1974, in B. M. McCormac (ed.), *Magnetospheric Physics*, Reidel, Dordrecht, Holland.
Boström, R.: 1975, in B. Hultqvist and L. Stenflo (eds.), *Physics of the Hot Plasma in the Magnetosphere*, Plenum Press, New York.
Carlqvist, P.: 1969, *Solar Phys.* **7**, 377.
Carlqvist, P.: 1972, *Cosmic Electrodyn.* **3**, 377.
Carlqvist, P.: 1973, TRITA-EPP-73-05, Royal Institute of Technology, Stockholm.
Carlqvist, P.: 1979a, in P. Palmadesso and K. Papadopoulos (eds.), *Wave Instabilities in Space Plasmas*, Proc. URSI Symp., Helsinki, August 1978, Reidel, Dordrecht, Holland.
Carlqvist, P.: 1979b, *Solar Phys.* **63**, 353.
Coakley, P., Hershkowitz, N., Hubbard, R., and Joyce, G.: 1978, *Phys. Rev. Letters* **40**, 230.
Coakley, P., Johnson, L., and Hershkowitz, N.: 1979, *Phys. Lett.* **70A**, 425.
DeGroot, J. S., Barnes, C., Walstead, A. E., and Buneman, O.: 1977, *Phys. Rev. Lett.* **38**, 1283.
Dreicer, H.: 1959, *Phys. Rev.* **115**, 238.
Dreicer, H.: 1960, *Phys. Rev.* **117**, 329.
Dungey, J. W.: 1961, *Phys. Rev. Letters* **6**, No. 2, 47.
Dungey, J. W.: 1964, *Space Sci. Rev.* **4**, 199.
Elphic, R. C., Russell, C. T., Slavin, J. A., and Brace, L. H.: 1979, 'Observations of the dayside ionopause and ionosphere of Venus', (submitted to *J. Geophys. Res.*).
Fälthammar, C.-G.: 1979, Invited lecture for the Workshop on Acceleration Mechanisms in Astrophysics, La Jolla, January 1979, TRITA-EPP-78-17, Royal Institute of Technology, Stockholm.
Fälthammar, C.-G., Akasofu, S.-I., and Alfvén, H.: 1978, *Nature* **275**, 185.
Giovanelli, R. G.: 1949, *Phil. Mag.* **40**, 206.
Gringauz, K. I.: 1980, *Proc. COSPAR Symp. on Planetary Magnetospheres*, Budapest, June 1980.
Gurnett, D. A.: 1972, E. R. Dyer (ed.), *Critical Problems of Magnetospheric Physics*, Proc. Joint COSPAR, IAGA, URSI Symp., Madrid, May 1972, IUCSTP, Washington, D.C.
Haerendel, G., Rieger, E., Valenzuela, A., Föppl, H., Stenbaek-Nielsen, H. C., and Wescott, E. M.: 1976, *European Programmes on Sounding-Rocket and Balloon Research in the Auroral Zone*, European Space Agency SP-115, Noordwijk, Holland.
Heikkila, W. J.: 1975, in B. Hultqvist and L. Stenflo (eds.), *Physics of the Hot Plasma in the Magnetosphere*, Plenum Publ. Co., New York.
Heikkila, W. J.: 1978, *Planet. Space Sci.* **26**, 121.
Heikkila, W. J. and Block, L. P.: 1977, *Geophysica* **14**, 2.
Hill, T. W.: 1975, *J. Geophys. Res.* **80**, 4689.
Hubbard, R. F. and Joyce, G.: 1979, *J. Geophys. Res.* **84**, 4297.
Ip, W.-H. and Mendis, D. A.: 1976, *Icarus* **29**, 147.
Kan, J. R., Akasofu, S.-I., and Lee, L. C.: 1980, in S.-I. Akasofu (ed.), *Dynamics of the Magnetosphere*, Proc. Chapman Conf., Los Alamos, October 1978, Reidel, Dordrecht, Holland.
Lennartsson, W.: 1979, TRITA-EPP-79-16, Royal Institute of Technology, Stockholm (accepted for publication in *Planetary Space Sci.*).
Leung, P., Wong, A. Y., and Quon, B. H.: 1979, *Phys. Fluids* **22**, No. 12.
Lutsenko, E. I., Sereda, N. D., and Kontsevoi, L. M.: 1975, *Zh. Tekh. Fiz.* **45**, 789 (*Sov. Phys. Tech. Phys.* **20**, 498).
Marklund, G.: 1978, TRITA-EPP-78-09, Royal Inst. of Technology, Stockholm.
Marklund, G.: 1979, *Nature* **277**, 370.
Marklund, G., Block, L., and Lindqvist, P.-A.: 1979, TRITA-EPP-79-22, Royal Institute of Technology, Stockholm.
Maynard, N. G.: 1978, *Geophys. Res. Lett.* **5**, 617.
Mendis, D. A.: 1978, *The Moon and the Planets* **18**, 361.
Mozer, F. S., Carlson, C. W., Hudson, M. K., Torbert, R. B., Parady, B., Yatteau, I., and Kelly, M. C.: 1977, *Phys. Rev. Letters* **38**, 292.
Murty, G. S.: 1961, *Ark. f. Fysik* **21**, 203.
Quon, B. H. and Wong, A. Y.: 1976, *Phys. Rev. Letters* **37**, 1393.
Russell, C. T., Elphic, R. C., and Slavin, J. A.: 1979, in B. Battrick (ed.), *Magnetospheric Boundary Layers*, Proc. Chapman Conf., Alpbach, June 1979, European Space Agency SP-148, Noordwijk, Holland.

Shawhan, S. D., Fälthammar, C.-G., and Block, L. P.: 1978, *J. Geophys. Res.* **83**, 1049
Smiddy, M., Kelley, M. C., Curke, W., Rich, F., Sagalyn, R., Shuman, B., Hays, R., and Lai, S.: 1977, *Geophys. Res. Lett.* **4**, 543.
Smith, E.: 1979, in B. Battrick (ed.), *Magnetospheric Boundary Layers*, Proc. Chapman Conf., Alpbach, June 1979, European Space Agency SP-148, Noordwijk, Holland.
Spicer, D. S.: 1977, *Solar Phys.* **53**, 305.
Torvén, S.: 1979, in P. Parmadesso and K. Papadopoulos (eds.), *Wave Instabilities in Space Plasmas*, Proc. URSI Symp., Helsinki, August 1978, Reidel, Dordrecht, Holland.
Torvén, S.: 1980, *Proc. Int. Conf. on Plasma Physics*, invited paper. Nagoya, April 1980 (in press).
Torvén, S. and Andersson, D.: 1979, *J. Phys. D. Appl. Phys.* **12**, 717.
Torvén, S. and Lindberg, L.: 1980, TRITA-EPP-80-02, Royal Institute of Technology, Stockholm.
von Engel, A. and Steenbeck, M.: 1932, *Elektrische Gasentladungen*, I, Springer, Berlin.
von Engel, A. and Steenbeck, M.: 1934, *Elektrische Gasentladungen*, II, Springer, Berlin.
Wescott, E. M., Stenbaek-Nielsen, H. C., Hallinan, T. J., Davis, T. N., and Peek, H. M.: 1976, *J. Geophys. Res.* **81**, 4495.
Wolf, R. A.: 1975, *Space Sci. Rev.* **17**, 537.

Chapter III

CE refers to Alfvén and Fälthammar (1963).
ESS refers to Alfvén and Arrhenius (1976).
Alfvén, H.: 1939, *Kungl. Sv. Vet.–Akademiens Handl.* (*3*), **18**, No. 3.
Alfvén, H.: 1955, *Tellus* 7, 50.
Alfvén, H.: 1978, *Astrophys. Space Sci.* **54**, 279.
Alfvén, H.: 1979, in B. Battrick (ed.) *Magnetospheric Boundary Layers*, Proc. Chapman Conf., Alpbach, June, 1979, European Space Agency SP-148, Noordwijk, Holland.
Alfvén, H. and Arrhenius, G.: 1976, *Evolution of the Solar System*, NASA Sp-345, Washington, D.C.
Alfvén, H. and Carlqvist, P.: 1967, *Solar Phys.* **1**, 220.
Alfvén, H. and Fälthammar, C.-G.: 1963, *Cosmical Electrodynamics*, Clarendon Press, Oxford.
Armstrong, J. C.: 1974, in B. M. McCormac (ed.), *Magnetospheric Physics*, Reidel, Dordrecht, Holland.
Baum, P. J., Bratenahl, and Kamin, G.: 1978, *Astrophys. J.* **226**, 286.
Block, L. P. and Fälthammar, C.-G.: 1976, *Ann. Geophys.* **32**, 161.
Boström, C. O., Sarris, E. T., Krimigis, S. M., Iijima, T., and Armstrong, T. P.: 1976, *EOS: Trans. AGU* **57**, 314.
Boström, R.: 1974, in B. M. McCormac (ed.), *Magnetospheric Physics*, Reidel, Dordrecht, Holland.
Burlaga, L. F., Lemaire, J. F., and Turner, J. M.: 1977, *J. Geophys. Res.* **82**, 3691.
Carlqvist, P.: 1969, *Solar Phys.* 7, 220.
Carlqvist, P.: 1973, TRITA-EPP-73-05, Royal Institute of Technology, Stockholm.
Carlqvist, P. and Alfvén, H.: 1980, *Astrophys. Space Sci.* (in press).
Cole, K.: 1974, *Planet. Space Sci.* **22**, 1975.
Cole, K.: 1976, *A.G.U. Chapman Symposium*, Yosemite, Feb. 1976.
Dungey, J. W.: 1961, *Phys. Rev. Letters* **6**, 47.
Dungey, J. W.: 1964, *Space Sci. Rev.* **4**, 199.
Elphic, R. C., Russell, C. T., Slavin, J. A., and Brace, L. H.: 1979, 'Observations of the dayside ionopause and ionosphere of Venus', (submitted to *J. Geophys. Res.*).
Gringauz, K. I.: 1980, *Proc. COSPAR Symp. on Planetary Magnetospheres*, Budapest, June 1980.
Hargrave, P. J. and Ryle, M.: 1974, *Monthly Notices Roy. Astr. Soc.* **166**, 305.
Heyvaerts, J.: 1979, *J. de Physique*, suppl. Colloque 7, **40**, C7-37.
Krimigis, S. M., Venkatesan, D., Barichello, J. C., and Sarris, E. T.: 1978, *Geophys. Res. Lett.* **5**, 961.
Lennartson, W.: 1977, *Astrophys. Space Sci.* **51**, 461.
Mende, S. B., Eather, R. H., and Weber, E. J.: 1980, in S.-I. Akasofu (ed.), *Dynamics of the Magnetosphere*, Proc. Chapman Conf., Los Alamos, October 1978, Reidel, Dordrecht, Holland.
Mendis, D. A.: 1978, *The Moon and the Planets* **18**, 361.
Moreton, G. E. and Severny, A. B.: 1968, *Solar Phys.* **3**, 282.
Ohyabu, N., Okomura, S., and Kawashima, N.: 1974, *Phys. Fluid* **17**, 2009.
Potemra, T. A.: 1979, *Rev. Geophys. Space Phys.* **17**, 640.

Potemra, T. A., Iijima, T., and Saflekos, N. A.: 1980, in S. I. Akasofu (ed.), *Dynamics of the Magnetosphere*, Proc. Chapman Conf., Los Alamos, October, 1978, Reidel, Dordrecht, Holland.
Rosenberg, R. L.: 1970, *Solar Phys.* **15**, 72.
Rosenberg, R. L. and Coleman, P. J., Jr.: 1969, *J. Geophys. Res.* **74**, 5611.
Rosenberg, R. L., Kivelson, M. G., and Hedgecock, P. C.: 1977, *J. Geophys. Res.* **82**, 1273.
Smith, E. J., Tsurutani, B. T., and Rosenberg, R. L.: 1977, *EOS Trans. AGU* **57**, 997.
Whipple, E. C., McIlwain, C. E., and Alfvén, H.: 1980, 'A three-ring circuit model of the magnetosphere', (submitted to *J. Geophys. Res.*).
Zmuda, A. J. and Armstrong, J. C.: 1974, *J. Geophys. Res.* **79**, 4611.

Chapter IV

CE refers to Alfvén and Fälthammar (1963).
ESS refers to Alfvén and Arrhenius (1976).
Alfvén, H.: 1942, *Stockholm Observatoriums Annaler* **14**(2), 3.
Alfvén, H.: 1950, *Tellus* **6**, 74.
Alfvén, H.: 1961, *Astrophys. J.* **133**, 1049.
Alfvén, H.: 1977, *Rev. Geophys. Space Phys.* **15**, 271.
Alfvén, H. and Arrhenius, G.: 1976, *Evolution of the Solar System*, (*ESS*), NASA SP-345, Washington, D.C.
Alfvén, H. and Fälthammar, C.-G.: 1963, *Cosmical Electrodynamics*, Int. Series of Monographs on Physics. Clarendon Press, Oxford.
Alfvén, H. and Klein, O.: 1962, *Ark. f. Fysik* **23**, 187.
Alfvén, H. and Lindberg, L.: 1974, *The Moon* **10**, 323.
Axnäs, I.: 1978a, *Astrophys. Space Sci.* **55**, 139.
Axnäs, I.: 1978b, 'Experimental Comparison of the Critical Velocity in Atom and Molecular Gases', TRITA-EPP-78-09, Royal Inst. of Technology, Stockholm.
Belcher, J. W. and Davis, J. R.: 1971, *J. Geophys. Res.* **76**, 3534.
Bennett, W. H.: 1934, *Phys. Rev.* **45**, 890.
Briggs, P. R., Armstrong, T. P., and Krimigis, S. M.: 1979, *Astrophys. J.* **228**, L83.
Danielsson, L.: 1970, *Phys. Fluids* **13**, 2288.
Danielsson, L.: 1973, *Astrophys. Space Sci.* **24**, 459.
De Forest, S. E.: 1972, *J. Geophys. Res.* **77**, 651.
Fälthammar, C.-G.: 1963, in B. Peters (ed.), *Proc. Int. School of Physics "Enrico Fermi"*, Course XIX, Academic Press, New York.
Fisk, L. A.: 1973, *Astrophys. J.* **224**, 1048.
Fowler, T. K. and Coensgen, F. H.: 1979, *9th Eur. Conf. Controlled Fusion and Plasma Physics*, Invited paper, Oxford, Sept, 1979.
Freeman, J. W., Fenner, M. A., Lindeman, R. A., Medrano, R., and Meister, J.: 1972, *Icarus* **16**, 328.
Gold, T. and Soter, S.: 1976, *Planet. Space Sci.* **24**, 45.
Goldenbaum, G. C., Irby, I. H., Chong, Y. P., and Hart, G.: 1979, *9th Eur. Conf. Controlled Fusion and Plasma Physics*, Contributed paper, Oxford, Sept. 1979.
Heikkila, W. J.: 1972, in E. R. Dyer (ed.), *Critical Problems of Magnetospheric Physics*, Proc. Joint COSPAR, IAGA, URSI Symp., Madrid, May 1972, IUCSTP, Washington, D.C.
Hill, J. R. and Mendis, D. A.: 1979, *Moon and the Planets* **21**, 3.
Hill, J. R. and Mendis, D. A.: 1980, 'Charged dust in outer planetary magnetospheres. II: Trajectories and spatial distribution', *Moon and the Planets* (in press).
Himmel, G., Möbius, E., and Piel, A.: 1976, *Z. Naturforsch.* **31a**, 934.
Himmel, G., Möbius, E., and Piel, A.: 1977, *Z. Naturforsch.* **32a**, 577.
Hirschberg, J.: 1973, *Rev. Geophys. Space Phys.* **11**, 115.
Hollweg, J. V.: 1975, *Rev. Geophys. Space Phys.* **13**, 263.
Kodomtsev, B. B.: 1965, *Plasma Turbulence*, Academic Press, New York.
Komori, A., Sato, N., Sugai, H., and Hatta, Y.: 1977, *Plasma Physics* **19**, 283.
Lehnert, B.: 1967, *Phys. Fluids* **10**, 2216.
Lehnert, B.: 1968, *Nuclear Fusion* **8**, 173.

Lehnert, B.: 1970, *Cosmic Electrodyn.* **1**, 397.
Lehnert, B.: 1975, *Nuclear Instr. and Methods* **129**, 31.
Lehnert, B.: 1976, *Astrophys. Space Sci.* **49**, 225.
Lehnert, B., Bergström, J., and Holmberg, S.: 1966, *Nuclear Fusion* **6**, 231.
Lindberg, L.: 1978, *Astrophys. Space Sci.* **55**, 203.
Lindberg, L. and Jacobsen, C.: 1961, *Astrophys. J.* **133**, 1043.
Lindberg, L. and Kristoferson, L.: 1971, *Cosmic Electrodyn.* **2**, 305.
Lindberg, L., Witalis, E., and Jacobsen, C. T.: 1960, *Nature* **185**, 452.
Lindeman, R. A., Vondrak, R. R., Freeman, J. W., and Snyder, C. W.: 1974, *J. Geophys. Res.* 79, 2287.
Mendis, D. A.: 1979, *Astrophys. Space Sci.* **65**, 5.
Mendis, D. A. and Ip, W.-H.: 1977, *Space Sci. Rev.* **20**, 145.
Möbius, E., Boswell, R. W., Piel, A., and Henry, D.: 1979b, *Geophys. Res. Lett.* **6**, 29.
Möbius, E., Piel, A., and Himmel, G.: 1979a, *Z. Naturforsch* **34a**, 405.
Morgan, D. L. and Hughes, V. W.: 1970, *Phys. Rev.* **D2**, 1389.
Murty, G. S.: 1962, *Ark f. Fysik* **21**, 203.
Petelski, E. F., Brenning, N., Ripken, H. W., Fahr, H. J., and Axnäs, I.: 1980, 'Enhanced interaction of the solar wind and the interstellar neutral gas by virtue of a critical velocity effect', *Astron. Astrophys.* (in press),
Piel, A., Möbius, E., and Himmel G.: 1980, *Astrophys. Space Sci.* (in press).
Raadu, M. A.: 1975, *TRITA-EPP-75-28*, Royal Inst. of Technology, Stockholm.
Raadu, M. A.: 1978, *Astrophys. Space Sci.* **55**, 125.
Raadu, M. A.: 1980, *Proc. Int. Workshop on Relation Between Laboratory and Space Plasma*, Tokyo, April 1980.
Reasoner, D. L., Lennartsson, W., Chappell, C. R., Rosen, A.(eds.): 1976, *Spacecraft Charging by Magnetospheric Plasmas*, Amer. Inst. of Aeronautics & Astronautics. N.Y., p. 89.
Risbeth, H., and Garriott, O. K.: 1969, *Introduction to Ionospheric Physics*, Academic Press, N.Y., p.1.
Rose, D. J. and Clark, M., Jr.: 1961, *Plasmas and Controlled Fusion*, MIT Press, Cambridge, Mass., p. 331.
Schlüter, V. A.: 1975, *Z. Naturforsch.* **12a**, 822.
Sherman, J. C.: 1973, *Astrophys. Space Sci.* **24**, 487.
Srnka, L. J.: 1977, *Phys. Earth Planet. Inter.* **14**, 321.
Srnka, L. J. and Merrill, R. B. (eds.): 1979, *Phys. Earth Planet. Inter.* **20**, Nos. 2–4.
Tanaka, U., Oya, K., Okuda, T., and Yamamoto, K.: 1972, *J. Physical Soc. Japan* **32**, No. 2.
Thompson, W. B.: 1978, *Astrophys. Space Sci.* **55**, 15.
Tsytovich, V. N.: 1977, *Theory of Turbulent Plasma*, Consultants Bureau, N.Y.
Varma, R. K.: 1978, *Astrophys. Space Sci.* **55**, 113.
Zmuda, A. J. and Armstrong, J. C.: 1974, *J. Geophys. Res.* **79**, 4611.
Zwickl, R. D., Roelof, E. C., Gold, R. E., Krimigis, S. M., and Armstrong, T. P.: 1978, *Astrophys. J.* **225**, 281.

Chapter V

ESS refers to Alfvén and Arrhenius (1976).
Alfvén, H.: 1942, *Stockholm Observatorium Annaler, I* **14**, No. 2.
Alfvén, H.: 1943, *Stockholm Observatorium Annaler, II* **14**, No. 5.
Alfvén, H.: 1946, *Stockholm Observatorium Annaler, III* **14**, No. 9.
Alfvén, H.: 1954, *On the Origin of the Solar System*, Oxford University Press, Oxford.
Alfvén, H.: 1963, *Astrophys. J.* **137**, 981.
Alfvén, H. and Arrhenius, G.: 1976, *Evolution of the Solar System*, NASA SP-345, Washington, D.C.
Alfvén, H. and Carlqvist, P.: 1978, *Astrophys. Space Sci.* **54**, 279.
Alfvén, H. and Wilcox, J. M.: 1962, *Astrophys. J.* **136**, 3.
De, B.: 1978, *Moon and the Planets* **18**, 339.
Cameron, A. G. W.: 1978, in S. F. Dermott (ed.), *Origin of the Solar System*, Proc. NATO Adv. Study Inst. Newcastle upon Tyne, March, 1976, Wiley, New York.

Gehrels, T. (ed.): 1978, *Protostars and Planets*, U. of Arizona Press, Tucson.
Goldreich, P. and Ward, W. E.: 1973, *Astrophys. J.* **183**, 1051.
Horedt, G. P.: 1976, *Astrophys. Space Sci.* **45**, 353.
Larson, R. B.: 1978, in T. Gehrels (ed.), *Protostars and Planets*, U. of Arizona Press, Tucson.
Prentice, A. J. R.: 1976, *Astron. Astrophys.* **50**, 59.
Reeves, H. (ed.): 1972, *Proc. Symp. on the Origin of the Solar System*, CNRS, Paris.
Safronov, V. S.: 1969, *Evolution of the Preplanetary Cloud and of the Earth and the Planets*, Nauka, Moscow.

Chapter VI

Alfvén, H. and Klein, O.: 1962, *Arkiv för fysik* **23**, 187.
Alfvén, H.: 1965, *Rev. Modern Physics* **37**, 4, 652.
Alfvén, H.: 1971, *Physics Today* **24**, (2), 28.
Alfvén, H.: 1977, in G. Ekspong and S. Nilsson (eds.), *Antinucleon–Nucleon Interactions* (Wenner-Gren Volume 29), Pergamon Press, Oxford.
Alfvén, H.: 1979a, *Astrophys. Space Sci.* **64**, 401.
Alfvén, H.: 1979b, *Astrophys. Space Sci.* **66**, 23.
Ambartzumian, V. A.: 1958, in R. Stoops (ed.), *Onzième Conseil de Physique Solvay* Bruxelles.
Arp, H.: 1966, *Science* **151**, 1214.
Arp, H.: 1977, in *Colloq. Int. No. 263*, CNRS, Paris.
Arp, H.: 1978, in L. V. Mirzoyan (ed.), *Problems of Physics and Evolution of the Universe*, Armenian Academy of Sciences.
Bonnevier, B.: 1978, Private communication; 1980, TRITA-EPP (in press).
Burbidge, G. R.: 1973, *Nature Physical Science* **246**, 17.
Burbidge, G. R.: 1979, *Nature* **282**, 451.
Burbidge, G. R., Jones, T. W., and O'Dell, S. T.: 1974, *Astrophys. J.* **193**, 43.
Burbidge, E. M., Smith, H. E., and Burbidge, G. R.: 1978, *Astrophys. J.* **219**, 400.
Carlqvist, P. and Laurent, B.: 1976a, *Nature* **260**, 225.
Carlqvist, P. and Laurent, B.: 1976b, *Nature* **262**, 822.
Colgate, S. A.: 1967, *Astrophys. J.* **150**.
de Vaucouleurs, G.: 1970, *Science* **167**, 1203.
de Vaucouleurs, G. and Bollinger, J.: 1979, *Astrophys. J.* **233**, 433.
Dohnanyi, J. S.: 1976 in H. Elasser and H. Fechtig (eds.), *Interplanetary Dust and Zodiacal Light*, Springer-Verlag.
Elvius, A.: 1970, in D. S. Evans (ed.), 'External Galaxies and Quasi Stellar Objects', *IAU Symp.* **44**, 306.
Fälthammar, C.-G., Akasofu, S.-I., and Alfvén, H.: 1978, *Nature* **275**, 185.
Fresnel, A.: 1826, Oeuvres complètes d'Augustin Fresnel (Publ. par de Senarmont, Verdet et Fresnel), Impr. Impériale, Paris.
Fry, J. N. and Peebles, P. J. E.: 1978, *Astrophys. J.* **221**, 19.
Gott, J., Gunn, E., Schramm, N., and Tinsley, M.: 1974, *Astrophys. J.* **194**, 543.
Groth, E. J. and Peebles, P. J. E.: 1977, *Astrophys. J.* **217**, 385.
Groth, E. J., Peebles, P. J. E., Seldner, M., and Soneira, R. M.: 1977, *Sci. American, Nov. 1977*, 76.
Jastrow, R.: 1978, *God and the Astronomers*, Norton & Co., New York.
Kristian, J., Sandage, A., and Westphal, J. A.: 1978, *Astrophys. J.* **221**, 383.
Lehnert, B.: 1977, *Astrophys. Space Sci.* **46**, 61.
Lehnert, B.: 1978, *Astrophys. Space Sci.* **53**, 459.
Malin, D. F. and Carter, D.: 1980, *Nature* **285**, 643.
Mathis, J. S.: 1967, *Astrophys. J.* **147**, 1050.
Nezhinskij, E. M.: 1972, in G. A. Chebatorev, E. I. Kazimirchak-Polonskaya, and B. G. Marsden (eds.), 'The Motion, Evolution of Orbits, and Origin of Comets', *IAU Symp.* **45**.
Omnès, R.: 1971, *Astron. Astrophys.* **10**, 228.
Pearson, T. J., Readhead, A. C. S., and Wilkinson, P. N.: 1980, *Astrophys. J.* **236**, 714.
Physica Scripta **17**, No. 3 (1978).

Readhead, A. C. S., Cohen, M. H., and Blandford, R. D.: 1978, *Nature* **272**, 131.
Rogers, S. and Thompson, W. B.: 1980, *Astrophys. Space Sci.* (in press).
Sandage, A., Tammann, G. A. and Hardy, E.: 1972, *Astrophys. J.* **172**, 253.
Sandage, A., Tammann, G. A. and Yahil, A.: 1979, *Astrophys. J.* **232**, 352.
Sanders, R. H.: 1970, *Astrophys. J.* **162**, 791.
Saslaw, W. C.: 1975, in A. Hayli (ed.), 'Dynamics of Stellar Systems', *IAU Symp.* **69**, 379.
Sciama, D. W.: 1971, *Modern Cosmology*, Cambridge University Press, Cambridge.
Seidl, F. G. P. and Cameron, A. G. W.: 1972, *Astrophys. Space Sci.* **15**, 44.
Smith, H. E., Junkkarinen, V. T., Spinrad, H., Grueff, G., and Vigotti, M.: 1979, *Astrophys. J.* **231**, 307.
Spitzer, L. and Saslaw, W. C.: 1966, *Astrophys. J.* **143**, 400.
Sofia, S. and Van Horn, H. M.: 1974, *Astrophys. J.* **194**, 593.
Sofia, S. and Van Horn, H. M.: 1975, *Ann. N.Y. Acad. Sci.* **262**, 197.
Steigman, G.: 1976a, *Nature* **262**, 821.
Steigman, G.: 1976b, *Annual Reviews*, 339.
Tarenghi, M., Chincarini, G., Rood, H. J., and Thompson, L. A.: 1980, *Astrophys. J.* **235**, 724.
Thompson, W. B.: 1978a, Private communication.
Thompson, W. B.: 1978b, *Astrophys. Space Sci.* **55**, 15.
Vincent, J. R.: 1976, TRITA-EPP-76-12, Royal Institute of Technology, Stockholm.
Weinberg, S.: 1977, *The First Three Minutes*, Basic Books, New York.
White, R. S., Ryan, J. M., Wilson, R. B., and Zych, A. D.: 1978, *Nature* **271**, 635.

INDEX

ASTROPHYSICS AND SPACE SCIENCE LIBRARY

Edited by

J. E. Blamont, R. L. F. Boyd, L. Goldberg, C. de Jager, Z. Kopal, G. H. Ludwig, R. Lüst, B. M. McCormac, H. E. Newell, L. I. Sedov, Z. Švestka, and W. de Graaff

1. C. de Jager (ed.), *The Solar Spectrum, Proceedings of the Symposium held at the University of Utrecht, 26–31 August, 1963.* 1965, XIV + 417 pp.
2. J. Orthner and H. Maseland (eds.), *Introduction to Solar Terrestrial Relations, Proceedings of the Summer School in Space Physics held in Alpbach, Austria, July 15–August 10, 1963 and Organized by the European Preparatory Commission for Space Research.* 1965, IX + 506 pp.
3. C. C. Chang and S. S. Huang (eds.), *Proceedings of the Plasma Space Science Symposium, held at the Catholic University of America, Washington, D.C., June 11–14, 1963.* 1965, IX + 377 pp.
4. Zdeněk Kopal, *An Introduction to the Study of the Moon.* 1966, XII + 464 pp.
5. B. M. McCormac (ed.), *Radiation Trapped in the Earth's Magnetic Field. Proceedings of the Advanced Study Institute, held at the Chr. Michelsen Institute, Bergen, Norway, August 16–September 3, 1965.* 1966, XII + 901 pp.
6. A. B. Underhill, *The Early Type Stars.* 1966, XII + 282 pp.
7. Jean Kovalevsky, *Introduction to Celestial Mechanics.* 1967, VIII + 427 pp.
8. Zdeněk Kopal and Constantine L. Goudas (eds.), *Measure of the Moon. Proceedings of the 2nd International Conference on Selenodesy and Lunar Topography, held in the University of Manchester, England, May 30–June 4, 1966.* 1967, XVIII + 479 pp.
9. J. G. Emming (ed.), *Electromagnetic Radiation in Space. Proceedings of the 3rd ESRO Summer School in Space Physics, held in Alpbach, Austria, from 19 July to 13 August, 1965.* 1968, VIII + 307 pp.
10. R. L. Carovillano, John F. McClay, and Henry R. Radoski (eds.), *Physics of the Magnetosphere, Based upon the Proceedings of the Conference held at Boston College, June 19–28, 1967.* 1968, X + 686 pp.
11. Syun-Ichi Akasofu, *Polar and Magnetospheric Substorms.* 1968, XVIII + 280 pp.
12. Peter M. Millman (ed.), *Meteorite Research. Proceedings of a Symposium on Meteorite Research, held in Vienna, Austria, 7–13 August, 1968.* 1969, XV + 941 pp.
13. Margherita Hack (ed.), *Mass Loss from Stars. Proceedings of the 2nd Trieste Colloquium on Astrophysics, 12–17 September, 1968.* 1969, XII + 345 pp.
14. N. D'Angelo (ed.), *Low-Frequency Waves and Irregularities in the Ionosphere. Proceedings of the 2nd ESRIN-ESLAB Symposium, held in Frascati, Italy, 23–27 September, 1968.* 1969, VII + 218 pp.
15. G. A. Partel (ed.), *Space Engineering. Proceedings of the 2nd International Conference on Space Engineering, held at the Fondazione Giorgio Cini, Isola di San Giorgio, Venice, Italy, May 7–10, 1969.* 1970, XI + 728 pp.
16. S. Fred Singer (ed.), *Manned Laboratories in Space. Second International Orbital Laboratory Symposium.* 1969, XIII + 133 pp.
17. B. M. McCormac (ed.), *Particles and Fields in the Magnetosphere. Symposium Organized by the Summer Advanced Study Institute, held at the University of California, Santa Barbara, Calif., August 4–15, 1969.* 1970, XI + 450 pp.
18. Jean-Claude Pecker, *Experimental Astronomy.* 1970, X + 105 pp.
19. V. Manno and D. E. Page (eds.), *Intercorrelated Satellite Observations related to Solar Events. Proceedings of the 3rd ESLAB/ESRIN Symposium held in Noordwijk, The Netherlands, September 16–19, 1969.* 1970, XVI + 627 pp.
20. L. Mansinha, D. E. Smylie, and A. E. Beck, *Earthquake Displacement Fields and the Rotation of the Earth, A NATO Advanced Study Institute Conference Organized by the Department of Geophysics, University of Western Ontario, London, Canada, June 22–28, 1969.* 1970, XI + 308 pp.
21. Jean-Claude Pecker, *Space Observatories.* 1970, XI + 120 pp.
22. L. N. Mavridis (ed.), *Structure and Evolution of the Galaxy. Proceedings of the NATO Advanced Study Institute, held in Athens, September 8–19, 1969.* 1971, VII + 312 pp.

23. A. Muller (ed.), *The Magellanic Clouds. A European Southern Observatory Presentation: Principal Prospects, Current Observational and Theoretical Approaches, and Prospects for Future Research, Based on the Symposium on the Magellanic Clouds, held in Santiago de Chile, March 1969, on the Occasion of the Dedication of the European Southern Observatory.* 1971, XII + 189 pp.
24. B. M. McCormac (ed.), *The Radiating Atmosphere. Proceedings of a Symposium Organized by the Summer Advanced Study Institute, held at Queen's University, Kingston, Ontario, August 3–14, 1970.* 1971, XI + 455 pp.
25. G. Fiocco (ed.), *Mesospheric Models and Related Experiments. Proceedings of the 4th ESRIN-ESLAB Symposium, held at Frascati, Italy, July 6–10, 1970.* 1971, VIII + 298 pp.
26. I. Atanasijević, *Selected Exercises in Galactic Astronomy.* 1971, XII + 144 pp.
27. C. J. Macris (ed.), *Physics of the Solar Corona. Proceedings of the NATO Advanced Study Institute on Physics of the Solar Corona, held at Cavouri-Vouliagmeni, Athens, Greece, 6–17 September 1970.* 1971, XII + 345 pp.
28. F. Delobeau, *The Environment of the Earth.* 1971, IX + 113 pp.
29. E. R. Dyer (general ed.), *Solar-Terrestrial Physics/1970. Proceedings of the International Symposium on Solar-Terrestrial Physics, held in Leningrad, U.S.S.R., 12–19 May 1970.* 1972, VIII + 938 pp.
30. V. Manno and J. Ring (eds.), *Infrared Detection Techniques for Space Research. Proceedings of the 5th ESLAB-ESRIN Symposium, held in Noordwijk, The Netherlands, June 8–11, 1971.* 1972, XII + 344 pp.
31. M. Lecar (ed.), *Gravitational N-Body Problem. Proceedings of IAU Colloquium No. 10, held in Cambridge, England, August 12–15, 1970.* 1972, XI + 441 pp.
32. B. M. McCormac (ed.), *Earth's Magnetospheric Processes. Proceedings of a Symposium Organized by the Summer Advanced Study Institute and Ninth ESRO Summer School, held in Cortina, Italy, August 30–September 10, 1971.* 1972, VIII + 417 pp.
33. Antonin Rükl, *Maps of Lunar Hemispheres.* 1972, V + 24 pp.
34. V. Kourganoff, *Introduction to the Physics of Stellar Interiors.* 1973, XI + 115 pp.
35. B. M. McCormac (ed.), *Physics and Chemistry of Upper Atmospheres. Proceedings of a Symposium Organized by the Summer Advanced Study Institute, held at the University of Orléans, France, July 31–August 11, 1972.* 1973, VIII + 389 pp.
36. J. D. Fernie (ed.), *Variable Stars in Globular Clusters and in Related Systems. Proceedings of the IAU Colloquium No. 21, held at the University of Toronto, Toronto, Canada, August 29–31, 1972.* 1973, IX + 234 pp.
37. R. J. L. Grard (ed.), *Photon and Particle Interaction with Surfaces in Space. Proceedings of the 6th ESLAB Symposium, held at Noordwijk, The Netherlands, 26–29 September, 1972.* 1973, XV + 577 pp.
38. Werner Israel (ed.), *Relativity, Astrophysics and Cosmology. Proceedings of the Summer School, held 14–26 August, 1972, at the BANFF Centre, BANFF, Alberta, Canada.* 1973, IX + 323 pp.
39. B. D. Tapley and V. Szebehely (eds.), *Recent Advances in Dynamical Astronomy. Proceedings of the NATO Advanced Study Institute in Dynamical Astronomy, held in Cortina d'Ampezzo, Italy, August 9–12, 1972.* 1973, XIII + 468 pp.
40. A. G. W. Cameron (ed.), *Cosmochemistry. Proceedings of the Symposium on Cosmochemistry, held at the Smithsonian Astrophysical Observatory, Cambridge, Mass., August 14–16, 1972.* 1973, X + 173 pp.
41. M. Golay, *Introduction to Astronomical Photometry.* 1974, IX + 364 pp.
42. D. E. Page (ed.), *Correlated Interplanetary and Magnetospheric Observations. Proceedings of the 7th ESLAB Symposium, held at Saulgau, W. Germany, 22–25 May, 1973.* 1974, XIV + 662 pp.
43. Riccardo Giacconi and Herbert Gursky (eds.), *X-Ray Astronomy.* 1974, X + 450 pp.
44. B. M. McCormac (ed.), *Magnetospheric Physics. Proceedings of the Advanced Summer Institute, held in Sheffield, U.K., August 1973.* 1974, VII + 399 pp.
45. C. B. Cosmovici (ed.), *Supernovae and Supernova Remnants. Proceedings of the International Conference on Supernovae, held in Lecce, Italy, May 7–11, 1973.* 1974, XVII + 387 pp.
46. A. P. Mitra, *Ionospheric Effects of Solar Flares.* 1974, XI + 294 pp.
47. S.-I. Akasofu, *Physics of Magnetospheric Substorms.* 1977, XVIII + 599 pp.

48. H. Gursky and R. Ruffini (eds.), *Neutron Stars, Black Holes and Binary X-Ray Sources.* 1975, XII + 441 pp.
49. Z. Švestka and P. Simon (eds.), *Catalog of Solar Particle Events 1955–1969. Prepared under the Auspices of Working Group 2 of the Inter-Union Commission on Solar-Terrestrial Physics.* 1975, IX + 428 pp.
50. Zdeněk Kopal and Robert W. Carder, *Mapping of the Moon.* 1974, VIII + 237 pp.
51. B. M. McCormac (ed.), *Atmospheres of Earth and the Planets. Proceedings of the Summer Advanced Study Institute, held at the University of Liège, Belgium, July 29–August 8, 1974.* 1975, VII + 454 pp.
52. V. Formisano (ed.), *The Magnetospheres of the Earth and Jupiter. Proceedings of the Neil Brice Memorial Symposium, held in Frascati, May 28–June 1, 1974.* 1975, XI + 485 pp.
53. R. Grant Athay, *The Solar Chromosphere and Corona: Quiet Sun.* 1976, XI + 504 pp.
54. C. de Jager and H. Nieuwenhuijzen (eds.), *Image Processing Techniques in Astronomy. Proceedings of a Conference, held in Utrecht on March 25–27, 1975.* XI + 418 pp.
55. N. C. Wickramasinghe and D. J. Morgan (eds.), *Solid State Astrophysics. Proceedings of a Symposium, held at the University College, Cardiff, Wales, 9–12 July 1974.* 1976, XII + 314 pp.
56. John Meaburn, *Detection and Spectrometry of Faint Light.* 1976, IX + 270 pp.
57. K. Knott and B. Battrick (eds.), *The Scientific Satellite Programme during the International Magnetospheric Study. Proceedings of the 10th ESLAB Symposium, held at Vienna, Austria, 10–13 June 1975.* 1976, XV + 464 pp.
58. B. M. McCormac (ed.), *Magnetospheric Particles and Fields. Proceedings of the Summer Advanced Study School, held in Graz, Austria, August 4–15, 1975.* 1976, VII + 331 pp.
59. B. S. P. Shen and M. Merker (eds.), *Spallation Nuclear Reactions and Their Applications.* 1976, VIII + 235 pp.
60. Walter S. Fitch (ed.), *Multiple Periodic Variable Stars. Proceedings of the International Astronomical Union Colloquium No. 29, held at Budapest, Hungary, 1–5 September 1976.* 1976, XIV + 348 pp.
61. J. J. Burger, A. Pedersen, and B. Battrick (eds.), *Atmospheric Physics from Spacelab. Proceedings of the 11th ESLAB Symposium, Organized by the Space Science Department of the European Space Agency, held at Frascati, Italy, 11–14 May 1976.* 1976, XX + 409 pp.
62. J. Derral Mulholland (ed.), *Scientific Applications of Lunar Laser Ranging. Proceedings of a Symposium held in Austin, Tex., U.S.A., 8–10 June, 1976.* 1977, XVII + 302 pp.
63. Giovanni G. Fazio (ed.), *Infrared and Submillimeter Astronomy. Proceedings of a Symposium held in Philadelphia, Penn., U.S.A., 8–10 June, 1976.* 1977, X + 226 pp.
64. C. Jaschek and G. A. Wilkins (eds.), *Compilation, Critical Evaluation and Distribution of Stellar Data. Proceedings of the International Astronomical Union Colloquium No. 35, held at Strasbourg, France, 19–21 August, 1976.* 1977, XIV + 316 pp.
65. M. Friedjung (ed.), *Novae and Related Stars. Proceedings of an International Conference held by the Institut d'Astrophysique, Paris, France, 7–9 September, 1976.* 1977, XIV + 228 pp.
66. David N. Schramm (ed.), *Supernovae. Proceedings of a Special IAU-Session on Supernovae held in Grenoble, France, 1 September, 1976.* 1977, X + 192 pp.
67. Jean Audouze (ed.), *CNO Isotopes in Astrophysics. Proceedings of a Special IAU Session held in Grenoble, France, 30 August, 1976.* 1977, XIII + 195 pp.
68. Z. Kopal, *Dynamics of Close Binary Systems,* XIII + 510 pp.
69. A. Bruzek and C. J. Durrant (eds.), *Illustrated Glossary for Solar and Solar-Terrestrial Physics.* 1977, XVIII + 204 pp.
70. H. van Woerden (ed.), *Topics in Interstellar Matter.* 1977, VIII + 295 pp.
71. M. A. Shea, D. F. Smart, and T. S. Wu (eds.), *Study of Travelling Interplanetary Phenomena.* 1977, XII + 439 pp.
72. V. Szebehely (ed.), *Dynamics of Planets and Satellites and Theories of Their Motion. Proceedings of IAU Colloquium No. 41, held in Cambridge, England, 17–19 August 1976.* 1978, XII + 375 pp.
73. James R. Wertz (ed.), *Spacecraft Attitude Determination and Control.* 1978, XVI + 858 pp.

74. Peter J. Palmadesso and K. Papadopoulos (eds.), *Wave Instabilities in Space Plasmas. Proceedings of a Symposium Organized Within the XIX URSI General Assembly held in Helsinki, Finland, July 31–August 8, 1978.* 1979, VII + 309 pp.
75. Bengt E. Westerlund (ed.), *Stars and Star Systems. Proceedings of the Fourth European Regional Meeting in Astronomy held in Uppsala, Sweden, 7–12 August, 1978.* 1979, XVIII + 264 pp.
76. Cornelis van Schooneveld (ed.), *Image Formation from Coherence Functions in Astronomy. Proceedings of IAU Colloquium No. 49 on the Formation of Images from Spatial Coherence Functions in Astronomy, held at Groningen, The Netherlands, 10–12 August 1978.* 1979, XII + 338 pp.
77. Zdeněk Kopal, *Language of the Stars. A Discourse on the Theory of the Light Changes of Eclipsing Variables.* 1979, VIII + 280 pp.
78. S.-I. Akasofu (ed.), *Dynamics of the Magnetosphere. Proceedings of the A.G.U. Chapman Conference 'Magnetospheric Substorms and Related Plasma Processes' held at Los Alamos Scientific Laboratory, N.M., U.S.A., October 9-13, 1978.* 1980, XII + 658 pp.
79. Paul S. Wesson, *Gravity, Particles, and Astrophysics. A Review of Modern Theories of Gravity and G-variability, and their Relation to Elementary Particle Physics and Astrophysics.* 1980, VIII + 188 pp.
80. Peter A. Shaver (ed.), *Radio Recombination Lines. Proceedings of a Workshop held in Ottawa, Ontario, Canada, August 24-25, 1979.* 1980, X + 284 pp.
81. Pier Luigi Bernacca and Remo Ruffini (eds.), *Astrophysics from Spacelab*, 1980, XI + 664 pp.
82. Hannes Alfvén, *Cosmic Plasma*, 1981, X + 160 pp.
83. Michael D. Papagiannis (ed.), *Strategies for the Search for Life in the Universe*, 1980, XVI + 254 pp.
84. H. Kikuchi (ed.), *Relation between Laboratory and Space Plasmas*, 1981, XII + 386 pp.
85. Peter van der Kamp, *Stellar Paths*, 1981, forthcoming.

GPSR Compliance

The European Union's (EU) General Product Safety Regulation (GPSR) is a set of rules that requires consumer products to be safe and our obligations to ensure this.

If you have any concerns about our products, you can contact us on ProductSafety@springernature.com

In case Publisher is established outside the EU, the EU authorized representative is:

Springer Nature Customer Service Center GmbH
Europaplatz 3
69115 Heidelberg, Germany

Batch number: 10614746

Printed by Printforce, the Netherlands